SpringerBriefs in Mathematics

SpringerBriefs present concise summaries of cutting-edge research and practical applications across a wide spectrum of fields. Featuring compact volumes of 50 to 125 pages, the series covers a range of content from professional to academic. Briefs are characterized by fast, global electronic dissemination, standard publishing contracts, standardized manuscript preparation and formatting guidelines, and expedited production schedules.

Typical topics might include:

- A timely report of state-of-the art techniques
- A bridge between new research results, as published in journal articles, and a contextual literature review
- A snapshot of a hot or emerging topic
- An in-depth case study
- A presentation of core concepts that students must understand in order to make independent contributions

SpringerBriefs in Mathematics showcases expositions in all areas of mathematics and applied mathematics. Manuscripts presenting new results or a single new result in a classical field, new field, or an emerging topic, applications, or bridges between new results and already published works, are encouraged. The series is intended for mathematicians and applied mathematicians. All works are peer-reviewed to meet the highest standards of scientific literature.

Titles from this series are indexed by Scopus, Web of Science, Mathematical Reviews, and zbMATH.

Boris Buffoni • John Toland

Connected Sets in Global Bifurcation Theory

 Springer

Boris Buffoni
Institute of Mathematics
École Polytechnique Fédérale de Lausanne
Lausanne, Switzerland

John Toland
Department of Mathematics
University of Bath
Bath, UK

ISSN 2191-8198   ISSN 2191-8201   (electronic)
SpringerBriefs in Mathematics
ISBN 978-3-031-87050-7   ISBN 978-3-031-87051-4   (eBook)
https://doi.org/10.1007/978-3-031-87051-4

This Springer imprint is published by the registered company Springer Nature Switzerland AG
The registered company address is: Gewerbestrasse 11, 6330 Cham, Switzerland

If disposing of this product, please recycle the paper.

Preface

The aim of this monograph is to explain from first principles and illustrate with examples how the point-set topology of metric spaces can help decide when connected and locally compact sets of solutions of nonlinear equations in Banach spaces are path-connected, Definition 4.1. A particular stimulus was the contrasting contributions to this question from two different theories, one topological and the other real-analytical, of global bifurcation.

- With minimal hypotheses on the smoothness of nonlinearities Rabinowitz [40, Thm. 1.3] used Leray-Schauder topological degree theory to prove (among other things) the existence of global, connected sets $\overline{\mathscr{C}}$ of solutions of nonlinear eigenvalue problems that bifurcate from a line of trivial solutions at an eigenvalue of odd multiplicity. But there is no mention of path-connectedness of $\overline{\mathscr{C}}$ even when the data is infinitely differentiable and the eigenvalue is simple.
- With the strong hypothesis that the nonlinearities are real-analytic, Dancer [14, 15] used the theory of real-analytical varieties to study the local structure of solution sets which lead (among other things) to the existence of global, path-connected sets of solutions of nonlinear eigenvalue problems that bifurcate from a line of trivial solutions at simple eigenvalues, and the paths are piecewise real-analytic. More generally, after stating [15, Thm. 4], he infers that when the nonlinearities are real-analytic the connected sets in Rabinowitz's theorem on bifurcations at eigenvalues of odd multiplicity can be replaced by path-connected sets.

This gives rise to the question:

> *What can be said about a global set of solutions which is connected and locally compact, but not path-connected?*

When solution sets are considered as connected, locally compact metric subspaces of the Banach spaces in which the equations are posed, the answer is well known in point-set topology, but the implications for global bifurcation may not have been fully recognised.

Since it is not difficult, see Chap. 7, to construct equations with infinitely differentiable nonlinearities for which $\overline{\mathscr{C}}$ is not path-connected (indeed $\overline{\mathscr{C}}$ may contain no non-trivial paths at all), it may appear that neither approach can be adapted to yield path-connectedness when the nonlinearities are infinitely differentiable but not real-analytic. However, classical point-set topology yields significant insights to this question and in Chap. 5 leads to some positive answers.

More generally, when the simple dichotomy that a connected set is either path-connected or not path-connected is added to Rabinowitz's famous dichotomy [40, Thm. 1.3,(i),(ii)], metric-space theory yields the following conclusions for the non-empty, connected, locally compact set $\overline{\mathscr{C}}$ in Theorem 1.3. (This $\overline{\mathscr{C}}$ in Chap. 1 below is denoted by $\mathscr{C}_\mu$ in [40, Thm. 1.3].)

(A) By Theorem 5.8, if $\overline{\mathscr{C}}$ is not path-connected it must contain a non-trivial compact, connected set of points at each of which the behaviour of $\overline{\mathscr{C}}$ is erratic, as typified by the points where the topologist's sine curve is not path-connected in Figs. 4.1 and 7.1. To reflect this behaviour such points will be called congestion points, see Definition 5.1.

(B) If $\overline{\mathscr{C}}$ has no congestion points, by Theorem 4.25, for $(\lambda, x) \in \overline{\mathscr{C}}$ and every $\epsilon > 0$ there is a closed neighbourhood V of (λ, x) such that diam $(V) < \epsilon$ and $\overline{\mathscr{C}} \setminus V$ has finitely many components.

(C) Even when $\overline{\mathscr{C}}$ has congestion points it may also have relatively open, connected subsets which have no congestion points. The behaviour of $\overline{\mathscr{C}}$ at points where such sets are zero distance from the set of congestion points can then be classified and interpreted for solution sets of the bifurcation problem (1.1), see Sects. 5.3 and 5.4. That all these behaviours can occur in bifurcation problems with infinitely differentiable nonlinearities is illustrated by examples in Remark 5.16 and Sect. 7.5.

(D) Examples in Chap. 7 show that in bifurcation problems a connected set $\overline{\mathscr{C}}$ may contain no non-trivial paths, even when the nonlinearity is infinitely differentiable expect at the point $(\lambda_0, 0)$ where bifurcation occurs.

(E) When studying connected sets of solutions of problem (1.1) in Banach spaces, the hypotheses of Theorem 5.5 and Corollaries 5.6 and 5.7, if verified by the equation at each solution (λ, x), ensure that the solution set $\overline{\mathscr{C}}$ is path-connected.

(F) If $F : \mathbb{R} \times X \to X$ is continuous on $\mathbb{R} \times X$ and real-analytic except at countably many points of $\mathbb{R} \times X$, every component of non-zero solutions of the nonlinear eigenvalue problem $F(\lambda, x) = 0$ is path-connected; see Theorem 5.32 for the details.

Prerequisites

The presentation is targeted at graduate students but should be accessible to anyone who has studied classical real analysis, metric spaces, and some linear functional analysis at undergraduate level. The main advice is to be wary of intuition about the properties of connected sets that may be true in linear spaces and especially in the plane, but are false in general metric spaces.

Key words and symbols are listed in the index which follows the references.

Acknowledgements
The authors are grateful to Geoffrey R. Burton for enlightening discussions, and to colleagues who drew our attention to [26, 27, 39], and its development in [2, 21, 24], that is relevant to Sects. 6.4 and 6.5. They also wish to thank Springer's Rémi Rodh for his patient support and wise advice.

Lausanne, Switzerland   Boris Buffoni
Bath, UK   John Toland

Declaration

Competing Interests The authors have no competing interests to declare that are relevant to the content of this manuscript.

Contents

Chapter 1
Introduction

Our aim is to extract new conclusions about global bifurcation from classical but elementary aspects of point-set topology, and in the process give a concise, self-contained account of the theory. To set the scene,

Sections 1.1 and 1.2 introduce the basic terminology of nonlinear eigenvalue problems and discuss the background to global bifurcation theory;
Sections 1.3 and 1.4 record the differences, both in hypotheses and conclusions, of two approaches to global bifurcation theory, one using topological, and the other real-analytic, techniques;
Section 1.5 summaries key facts about connected, locally compact sets that are not path-connected and Sect. 1.6 reviews their implications for path-connectedness in global bifurcation theory.

After that the material is organised as follows.

Chapter 2 introduces notation and records, without proofs, results from set theory.

Chapter 3 recalls familiar terminology from metric space theory [37, 49, 51], including Baire category, metric subspaces, compactness, local compactness, para-compactness, separability, separations and connectedness, components and composants, simple chains, and convergence of sequences of sets. It is noteworthy that Theorem 3.53 is the separation result [40, Lemma 1.1] that plays a central role in Rabinowitz's global bifurcation theory.

Chapter 4 concerns connectedness and path-connectedness, and the difference between local connectedness and weak local connectedness at a point.

A connected set which is locally connected and locally compact is called a generalised Peano continuum and, by Theorem 4.25, if $p \in M$ where M is a generalised Peano continuum, there are closed neighbourhoods V of p with arbitrarily small diameter for which $M \setminus V$ has finitely many components, and M is path-connected by Theorem 4.12.

© The Author(s), under exclusive license to Springer Nature Switzerland AG 2025
B. Buffoni, J. Toland, *Connected Sets in Global Bifurcation Theory*, SpringerBriefs
in Mathematics, https://doi.org/10.1007/978-3-031-87051-4_1

When combined with Lemma 4.23, Theorem 4.12 yields the key observation that a connected and locally compact metric space which is not path-connected contains a point at which it is not weakly locally connected.

In Chap. 5, a point x at which M is not weakly locally connected is referred to as a congestion point because by Theorem 5.2, if M is connected the boundaries of arbitrarily small neighbourhoods of congestion points have infinitely many components. Moreover, if M is connected and locally compact, by Theorem 5.8 all balls which are centred at a congestion point with sufficiently small radii have infinitely many components.

In Sect. 5.1, which concerns connected subsets of Banach spaces, Corollary 5.7 leads to the conclusion, Theorem 5.15, that if the solutions $x \in X$ of (1.1a) are isolated for each fixed $\lambda \in \mathbb{R}$, the global connected set of solutions is path-connected in $\mathbb{R} \times X$.

Section 5.2 describes the complex behaviour of a connected, locally compact metric space in arbitrarily small neighbourhoods of a congestion point and in Theorem 5.8 shows that no component of congestion points is a singleton. This leads to refinements that may be added to Rabinowitz's classical dichotomy [40, Thm.1.3 (i),(ii)] in global bifurcation theory. In Sects. 5.3 and 5.4 these refinements are characterised by the local structure of the set of the non-trivial components of congestion points.

In Sect. 5.5, topological and real-analytic theories are merged with point-set topology to treat problem (1.1) when R is compact, continuous everywhere and real-analytic except at points of a totally disconnected set[1]. As a consequence, if R is real-analytic except at countably many points, by Theorem 5.32 the components of the set of non-trivial solutions of (1.1) that bifurcate from eigenvalues of odd multiplicity [40] are path-connected.

Chapter 6 concerns decomposable, indecomposable and hereditarily indecomposable, compact, connected sets, and some historical background is in order.

A continuum is a non-empty compact, connected set and a continuum is decomposable if it is the union of a pair of distinct proper sub-continua. For example, a path $P = \{f(t) : t \in [0, 1]\}$, where $f : [0, 1] \to \mathbb{R}^2$ is continuous and in addition $f^{-1}(\{f(0)\})$ and $f^{-1}(\{f(1)\})$ have one point each, is decomposable because, for all $\epsilon \in (0, 1)$,

$$P = \{f(t) : t \in [0, \epsilon]\} \cup \{f(t) : t \in [\epsilon, 1]\}$$

and $\{f(t) : t \in [0, \epsilon]\} \neq P \neq \{f(t) : t \in [\epsilon, 1]\}$. This is a special case of Lemma 6.8.

In 1910, L. E. J. Brouwer [6] gave the first example of a continuum which is not decomposable and 1922 Knaster [25] found a more exotic continuum: a non-degenerate one which itself and all its sub-continua are indecomposable.

---

[1] In [15] Dancer remarked in passing that his theory might be extended to equations with operators that are real-analytic everywhere, except at points of a totally disconnected set.

Continua with the latter property are called hereditarily indecomposable, and their existence means that there are compact, connected sets which contain no nontrivial paths, since a non-trivial path would be a decomposable sub-continuum. Knaster's set was later named the pseudo-arc by Moise [34].

In 1930 Mazurkiewicz [33] showed that far from being rare most continua in the plane are hereditarily indecomposable in the sense that they are second Baire category in the Hausdorff metric space of continua in $\mathbb{R}^2$. In 1951 Bing [4, Thm. 2] extended this result to a more general setting.

Chapter 6 begins by highlighting properties of sets in $\mathbb{R}^2$ that facilitate the proofs of existence of indecomposable continua in Sect. 6.3; Knaster's theorem in Sect. 6.4; and Mazurkiewicz's result on Baire category in Sect. 6.5.

Chapter 7 is devoted to the construction of examples in which the connected solution sets of problem (1.1) predicted by topological bifurcation theory are connected and locally compact but contain no non-trivial paths, although the hypotheses of [40] are satisfied.

These examples and the results of Mazurkiewicz and Bing mean that connected, locally compact sets of solutions arising in applications should not be regarded casually as if they are path-connected. This is where results such as Theorems 5.15 and 5.32 may come into play.

1.1 Nonlinear Eigenvalue Problems

Definition 1.1 When X and Y are Banach spaces, a mapping $F : X \to Y$ is compact if $\overline{F(U)}$ is compact in Y when U is bounded in X. When F is linear and compact it is automatically continuous. □

With $\mathbb{R}$ denoting the real numbers and X a real Banach space, consider the nonlinear equation for $(\lambda, x) \in \mathbb{R} \times X$,

$$x = \lambda L x + R(\lambda, x), \tag{1.1a}$$

where $L : X \to X$ is linear and compact, $R : \mathbb{R} \times X \to X$ is continuous and compact, $R(\lambda, 0) = 0$ for all $\lambda \in \mathbb{R}$ and

$$\frac{\|R(\lambda, x)\|}{\|x\|} \to 0 \text{ as } 0 \neq \|x\| \to 0 \text{ uniformly for } \lambda \text{ in bounded sets.} \tag{1.1b}$$

Let $\mathcal{T} = \{(\lambda, 0) : \lambda \in \mathbb{R}\}$ denote the line of trivial solutions of (1.1a) and $\mathcal{S}$ the solutions of (1.1a) that are not in $\mathcal{T}$. Then according to Krasnosel'skii [28, p. 181], λ_0 is a bifurcation point and $\mathcal{S}$ bifurcates at $(\lambda_0, 0)$ if there exists

$$(\lambda_j, x_j) \in \mathcal{S}, \quad j \in \mathbb{N}, \text{ such that } (\lambda_j, x_j) \to (\lambda_0, 0) \text{ in } \mathbb{R} \times X \text{ as } j \to \infty.$$

System (1.1) is a nonlinear eigenvalue problem and its bifurcation theory seeks sufficient conditions for λ_0 to be a bifurcation point and then studies the behaviour of sets of solutions in $\mathcal{S}$ that bifurcate from $\mathcal{T}$ at $(\lambda_0, 0)$. Dividing (1.1a) by $\|x_j\|$ and letting $j \to \infty$ implies that under the above hypotheses λ_0 is a bifurcation point only if $\lambda_0 \neq 0$ and λ_0^{-1} is an eigenvalue of L.

Definition 1.2 The reciprocal λ_0 of a non-zero eigenvalue of L is called a characteristic value of L and $\dim \left(\cup_{k \in \mathbb{N}} \ker(I - \lambda_0 L)^k \subset X \right)$ is the multiplicity of λ_0. A characteristic value is simple if its multiplicity is 1.   □

Since L is compact and X is a Banach space, all characteristic values of L have finite multiplicity, are isolated and bounded away from zero, and its set of characteristic values may be unbounded if X is infinite dimensional [5].

1.2   Global Bifurcation

The trivial example, $x = \lambda x + i|x|x$, $(\lambda, x) \in \mathbb{R} \times \mathbb{C}$, in which $\lambda_0 = 1$ is a characteristic value of multiplicity 2, shows that not all characteristic values are bifurcation points. Nevertheless Krasnosel'skii [28, Thm. 2.1, p. 196] used topological degree theory [30] to prove that every characteristic value of odd multiplicity is a bifurcation point. However, although his definition [28, p. 187] of a "branch" of solutions is global,[2] it is not required to be connected.

Rabinowitz [40] revealed the full power of the method when he invoked a separation result, Theorem 3.53 below, to prove under Krasnosel'skii's hypotheses the existence of connected sets of solutions that are either unbounded or their closures intersect the line of trivial solutions at an even number of distinct bifurcation points of odd multiplicity. The existence of global connected sets of non-trivial solutions satisfying this dichotomy now typify global bifurcation theory.

Significantly degree-theory methods work under minimal hypotheses on the smoothness of the nonlinear operators in the equation but they say nothing about path-connectedness even when the operators are smooth.

By contrast, when R in (1.1) is real-analytic (Sect. 1.4) the theory of [14] and [15] leads among other things to the existence of a global, piecewise-real-analytic curve of non-trivial solutions that bifurcates from simple characteristic values.

---

[2] In 1960, Krasovskii [29] used the theory of [28, §V.2] to obtain the existence of large amplitude Stokes waves.

1.3 Topological Theory

Theorem 1.3 (a) is a special case of Rabinowitz's celebrated theory of global bifurcation at characteristic values of odd multiplicity. As in [40, p. 488] let $\mathscr{S} = \overline{S}$, the closure of the set of non-trivial solutions of (1.1a).

Theorem 1.3 *For problem* (1.1) *suppose the hypotheses of Sect. 1.1 are satisfied and λ_0 is a characteristic value of L of odd multiplicity.*

(a) *Then there is a set $\mathscr{C}$ in S such that $\overline{\mathscr{C}}$ is a component of $\mathscr{S}$ with $(\lambda_0, 0) \in \overline{\mathscr{C}}$ and either $\overline{\mathscr{C}}$ is unbounded in $\mathbb{R} \times X$ or there exists $(\lambda_1, 0) \in \overline{\mathscr{C}}$ where $\lambda_1 \neq \lambda_0$ and λ_1 is a characteristic value of odd multiplicity of L.*
   With $\mathscr{C}_\mu$ as defined in [40, Thm. 1.3], $\overline{\mathscr{C}} = \mathscr{C}_\mu$ and $\mathscr{C} = \mathscr{C}_\mu \setminus \mathcal{T}$.

In particular, part (a) holds if λ_0 is simple. But if λ_0 is simple, part (b) says more about the set of solutions in a neighbourhood of $(\lambda_0, 0)$ when R satisfies additional hypotheses in a neighbourhood of $(\lambda_0, 0)$.

(b)(i) *If λ_0 is simple, R is continuously differentiable and $\partial_{\lambda x} R$ is continuous in a neighbourhood of $(\lambda_0, 0)$, there is an open neighbourhood U of $(\lambda_0, 0)$ and a homeomorphism $\gamma : (-1, 1) \to \overline{\mathscr{C}} \cap U$ such that $\gamma(0) = (\lambda_0, 0)$ and $U \cap S = \{\gamma(s) : s \in (-1, 0) \cup (0, 1)\}$.*

(b)(ii) *If λ_0 is simple and R is twice continuously differentiable in a neighbourhood of $(\lambda_0, 0)$ there is an open neighbourhood U of $(\lambda_0, 0)$ and a continuously differentiable homeomorphism $\gamma : (-1, 1) \to \overline{\mathscr{C}} \cap U$ such that $\gamma(0) = (\lambda_0, 0)$ and $U \cap S = \{\gamma(s) : s \in (-1, 0) \cup (0, 1)\}$.*

Part (b)(i) is a special case of [10, Thm. 1.7]. (In their notations, αx_0 is the projection along Z on $N(F_x(0, 0))$ of $(\varphi(\alpha), \alpha x_0 + \alpha \psi(\alpha)) \in F^{-1}(0) \cap U$ with $\psi(\alpha) \in Z$, and their parameter α can therefore be continuously recovered.) Part (b)(ii) is a corollary of [9, Ch. 5, Thm. 5.1], which shows $\mathscr{S} \cap U = (\mathcal{S} \cap U) \cup \{(\lambda_0, 0)\}$ is a non-self-intersecting C^1-curve in $\mathbb{R} \times X$. These proofs, which depend on the implicit function theorem, do not detect the global nature of bifurcation in part (a).

1.4 Real-Analytic Theory

A function or operator is real-analytic at a point if it is the sum of its Taylor series uniformly in a ball about the point, see Definition 5.19 and Remark 5.20. When L and R are compact, $R : \mathbb{R} \times X \to X$ is a real-analytic operator and λ_0 is a simple characteristic value of L, by Dancer's theory in [8, § 2.1] the global connected set of solutions of (1.1), which by Theorem 1.3 bifurcates from the trivial solutions at $(\lambda_0, 0)$, contains a global path

$$\mathscr{K} = \{(\Lambda(s), \kappa(s)) : s \in [0, \infty)\} \subset \mathbb{R} \times X,$$

where $s \mapsto \big(\Lambda(s), \kappa(s)\big)$ is continuous, with the following properties.

(i) $\Lambda(0) = \lambda_0 \in \mathbb{R}$, $\kappa(0) = 0 \in X$ and $\mathcal{K} \setminus \{(\lambda_0, 0)\}$ is a real-analytic curve in a neighbourhood of $(\lambda_0, 0)$, [7, page 6 and Thm. 9.1.1].

(ii) $\mathcal{K}$ is either unbounded or forms a closed loop in $\mathbb{R} \times X$.

(iii) For each $s^* \in (0, \infty)$ there exists $\rho^* : (-1, 1) \to [0, \infty)$ (a local re-parametrisation) which is continuous, injective, and

$$\rho^*(0) = s^* \quad \text{and} \quad t \mapsto (\Lambda(\rho^*(t)), \kappa(\rho^*(t))) \text{ is analytic on } t \in (-1, 1).$$

This does not imply that $\mathcal{K}$ is smooth ($\sigma : (-1, 1) \to \mathbb{R}^2$ defined by $\sigma(t) = (t^2, t^3)$ is real-analytic but the path it describes has a cusp), nor does it preclude the possibility of secondary bifurcation points on $\mathcal{K}$. Indeed, since (Λ, κ) is not required to be globally injective, self-intersection of $\mathcal{K}$ (as in a figure eight) is not ruled out.

(iv) Secondary bifurcation points and points where the bifurcating branch $\mathcal{K}$ is not smooth, if any, are isolated.

(v) This path is unique in the sense that it has a pre-determined continuation through secondary bifurcation points, and even through points where it intersects higher-dimensional manifolds of solutions.

In the example of (1.1) in Sect. 7.3, the bifurcation point $\lambda_0 = 1$ is simple and $R : \mathbb{R}^2 \to \mathbb{R}$ is a C^1-function which is infinitely differentiable, but not real-analytic, on $\mathbb{R}^2 \setminus \{(1, 0)\}$, and the global set $\overline{\mathscr{C}}$ in Theorem 1.3(a) contains no paths.

In the example of (1.1) in Sect. 7.4, the bifurcation point $\lambda_0 = 1$ is simple and $R : \mathbb{R}^2 \to \mathbb{R}$ is everywhere infinitely differentiable but not real-analytic, and in Theorem 1.3(b) $\overline{\mathscr{C}}$ is the union of three disjoint connected sets,

$$\overline{\mathscr{C}} = \mathscr{C}^- \cup \mathscr{L} \cup \mathscr{C}^+ \text{ where } (1, 0) \in \mathscr{L}, \text{ a real-analytic curve,} \qquad (1.2)$$

but $\mathscr{C}^\pm$ are unbounded connected sets containing no non-trivial paths.

1.5   Congestion Points

A metric space is said to be weakly locally connected at x if every neighbourhood of x contains a connected neighbourhood[3] of x, and a point where it is not weakly locally connected will be called a congestion point. Congestion points are important because, by Theorem 4.12 and Lemma 4.23, a connected, locally compact metric space which is not path-connected must contain a congestion point.

---

[3] This is different from local connectedness at a point, see Remark 4.22.

For example, since $\overline{\mathscr{C}}$ in Theorem 1.3 is connected and locally compact, if it is not path-connected its set of congestion points $\mathscr{N}(\overline{\mathscr{C}})$ is non-empty and when (λ, x) is a congestion point there exists $\epsilon_0 > 0$ such that, when $B_\epsilon(\lambda, x)$ is the open ball centred at (λ, x) with radius ϵ in $\mathbb{R} \times X$, for $\epsilon \in (0, \epsilon_0)$:

(i) the boundary in $\overline{\mathscr{C}}$ of every neighbourhood U of (λ, x) with diam $(U) < \epsilon$ has infinitely many components;
(ii) there is a sequence $\{K_j\}$ of components of $B_\epsilon(\lambda, x) \cap \overline{\mathscr{C}}$ such that

$$(\lambda, x) \notin K_j, \quad \overline{K_j} \cap \partial B_\epsilon(\lambda, x) \neq \emptyset, \quad j \in \mathbb{N},$$

and dist $\big((\lambda, x), K_j\big) \to 0$ as $j \to \infty$;
(iii) No component of $\mathscr{N}(\overline{\mathscr{C}})$ is a singleton.

A singleton is a set which has exactly one point.

Figure 7.1 illustrates the solution set $\overline{\mathscr{C}}$ of an explicit example of problem (1.1) in which there are two components of congestion points of $\overline{\mathscr{C}}$.

1.6   In Conclusion

Because of the examples in Chap. 7 where $\mathscr{C}$ has no non-trivial paths, there may seem to be no hope for a global bifurcation theory that guarantees path-connectedness of solution sets of problem (1.1) except when R is real-analytic.

However, Theorem 5.15 shows that path-connectedness is guaranteed by the theory of congestion points if the solution set has additional properties. Moreover, Theorem 5.32 combines the topological degree theory of [40], the theory of congestion points in Theorem 5.8, and the local structure of real-analytic functions in [15], to obtain a path-connected global bifurcation theory of problem (1.1) for the class of continuous and compact functions $R : \mathbb{R} \times X \to X$ which are real-analytic except at countably many points.

Chapter 2
Set Theory Foundations

In what follows, $\mathbb{Q}$, $\mathbb{Z}$ and $\mathbb{N}$ respectively denote the rational numbers, the integers, the natural numbers (positive integers not including zero), and $\emptyset$ is the empty set. If $\emptyset \neq A \subset B \neq A$, then A is a proper subset of B. The following, which is fundamental in set theory, will be assumed throughout.

The Axiom of Choice For an arbitrary non-empty family $\{P_\alpha : \alpha \in \mathscr{A}\}$ of non-empty sets there exists a set P which contains one and only one element from each P_α, $\alpha \in \mathscr{A}$. □

In set theory this seemingly obvious assertion is equivalent to several other statements which are less obvious, but play an important role in many branches of mathematics. To describe without proofs one of these statements, further definitions are needed.

For a non-empty set S, a subset R of $S \times S$ defines a relation $\sim$ on S by saying x is related to y, written $x \sim y$, if $(x, y) \in R$.

Definition 2.1 A relation $\sim$ on S is

   (i) reflexive if $x \sim x$ for all $x \in S$;
  (ii) symmetric if $x \sim y$ if and only if $y \sim x$ for all $x, y \in S$;
 (iii) transitive if $x \sim y$ and $y \sim z$ implies $x \sim z$ for all $x, y, z \in S$;
 (iv) anti-symmetric if $x \sim y$ and $y \sim x$ implies $x = y$ for all $x, y \in S$.

An equivalence relation is reflexive, symmetric and transitive while a partial ordering is a relation which is reflexive, anti-symmetric and transitive, in which case $(S, \sim)$ is called a partially ordered set. When $\sim$ is an equivalence relation the set $\{y \in S : x \sim y\}$, $x \in S$, is called the equivalence class of $\sim$ which contains x. □

Remark 2.2 Obviously, $(\mathbb{R}, \leqslant)$, where $\leqslant$ is the usual ordering of real numbers, is a partially ordered set, and if $\mathscr{P}(X)$ is the set of subsets of X, $(\mathscr{P}(X), \subset)$ is a partially ordered where $\subset$ denotes set inclusion. □

© The Author(s), under exclusive license to Springer Nature Switzerland AG 2025
B. Buffoni, J. Toland, *Connected Sets in Global Bifurcation Theory*, SpringerBriefs in Mathematics, https://doi.org/10.1007/978-3-031-87051-4_2

Definition 2.3 When $(S, \preceq)$ is a partially ordered set and $\emptyset \neq T \subset S$,

(a) $y \in T$ is a minimum element of T if $y \preceq x$ for all $x \in T$;
(b) T is well-ordered if every non-empty subset of T has a minimum element;
(c) T is totally ordered if either $x \preceq y$ or $y \preceq x$, for all $x, y \in T$.   □

Note that if T has a minimum element it is unique because $\preceq$ is anti-symmetric, and clearly every well-ordered set is totally ordered.

The following statement is equivalent to the Axiom of Choice.

The Well-Ordering Principle (Zermelo's Theorem) For any non-empty set S, there is a partial ordering $\preceq$ such that $(S, \preceq)$ is a well-ordered set.   □

The well-ordering principle does not describe the relation $\preceq$ that makes S well-ordered, but its existence is all that is needed in the proof of Theorem 3.14.

Definition 2.4 A set is countable if there is an injective function from it to the natural numbers; a set is uncountable otherwise.   □

Remark 2.5 If $A \subset B$ and B is countable, A is countable, and B is uncountable if A is uncountable; the union of a countable collection of countable sets is countable; the rationals $\mathbb{Q}$ are countable but an interval $[a, b] \subset \mathbb{R}$, $a < b$, is uncountable, and hence $\mathbb{R}$ is uncountable.   □

Chapter 3
Metric Spaces

3.1 Notation

Throughout, (M, d), $M \neq \emptyset$, is a metric space,

$$B_\epsilon(x) = \{y \in M : d(x, y) < \epsilon\}, \quad \epsilon > 0,$$

is the open ball of radius ϵ centred at $x \in M$ and a neighbourhood of x is any set which contains an open ball centred at x. For a non-empty subset $A \subset M$ and $x \in M$,

$$\text{dist}\,(x, A) = \inf\{d(x, y) : y \in A\} \text{ is the distance from } x \text{ to } A,$$

and for $\epsilon > 0$

$$N_\epsilon(A) = \{x \in M : \text{dist}\,(x, A) < \epsilon\} \text{ is the } \epsilon\text{-neighbourhood of } A. \tag{3.1a}$$

For $A \subset M$, let $\overline{A}$, A° and $\partial A = \overline{A} \setminus A^\circ$ denote the closure, interior and boundary of A. Note that

$$\partial B_\epsilon(x) \subset \{y \in M : d(x, y) = \epsilon\} = S_\epsilon(x),$$

but equality does not always hold, see Theorem 4.26. For $A \subset M$, the complement of A in M is

$$cA = M \setminus A = \{x \in M : x \notin A\}, \tag{3.1b}$$

and hence

$$c(\overline{A}) = (cA)^\circ \text{ and } c(A^\circ) = \overline{cA}. \tag{3.1c}$$

© The Author(s), under exclusive license to Springer Nature Switzerland AG 2025
B. Buffoni, J. Toland, *Connected Sets in Global Bifurcation Theory*, SpringerBriefs
in Mathematics, https://doi.org/10.1007/978-3-031-87051-4_3

If (M_1, d_1) and (M_2, d_2) are metric spaces, a product metric d is defined on the Cartesian product $M_1 \times M_2$ by

$$d\big((x_1, y_1), (x_2, y_2)\big) = \sqrt{d_1(x_1, x_2)^2 + d_2(y_1, y_2)^2}. \qquad (3.1\mathrm{d})$$

There are other product metrics, but (3.1d) suffices for what follows.

Definition 3.1 A collection $\mathscr{G}_{\mathscr{A}} = \{G_\alpha : \alpha \in \mathscr{A}\}$ of subsets of M is a cover of M if $M \subset \cup_{\alpha \in \mathscr{A}} G_\alpha$; the cover $\mathscr{G}_{\mathscr{A}}$ is finite if $\mathscr{A}$ is finite and $\mathscr{G}_{\mathscr{A}}$ is open if all the G_α are open. Another cover $\mathscr{H}_{\mathscr{B}} = \{H_\beta : \beta \in \mathscr{B}\}$ of M is a sub-cover of $\mathscr{G}_{\mathscr{A}}$ if $\mathscr{H}_{\mathscr{B}} \subset \mathscr{G}_{\mathscr{A}}$. □

Definition 3.2 A metric space M is

(i) compact if every open cover of M has a finite sub-cover, equivalently, any sequence in M has a convergent subsequence;
(ii) locally compact if every point of M has a compact neighbourhood;
(iii) σ-compact if M is a union of countably many compact sets;
(iv) complete if every Cauchy sequence is convergent, where a sequence $\{x_k\}$ is Cauchy if for $\epsilon > 0$ there exists $K \in \mathbb{N}$ such that $k, k' \geqslant K$ implies $d(x_k, x_{k'}) < \epsilon$. □

Definition 3.3 A set $D \subset M$ is dense in M if $\overline{D} = M$, and D is nowhere dense in M if $\big(\overline{D}\big)^\circ = \emptyset$. □

Lemma 3.4 *If U is open and dense in M, then cU is nowhere dense in M and if V is nowhere dense in M, $c(\overline{V})$ is open and dense in M.*

Proof If U is open and dense, $\emptyset = c(\overline{U}) = (cU)^\circ = (\overline{cU})^\circ$ by (3.1c), and if V is nowhere dense, $c(\overline{V})$ is open and $M = c((\overline{V})^\circ) = \overline{c(\overline{V})}$. □

Definition 3.5 M is separable if it has a countable dense subset. □

It is easily seen that compact sets, and hence σ-compact sets, are separable. See Theorem 3.28 for a result about locally compact spaces.

Definition 3.6 A family B of open subsets of M is called a base for M if, for every $x \in M$ and every open set G with $x \in G$, there exists $B \in B$ with $x \in B \subset G$. □

Remark 3.7 If M is separable, $M = \overline{D}$ where D is countable, and so

$$B = \big\{B_{1/k}(z) : z \in D, \ k \in \mathbb{N}\big\}$$

is countable by Remark 2.5. Therefore B is a countable base for M since, for $x \in G$ where G is open, there exists $B \in B$ with $x \in B \subset G$. □

3.2 Baire Category

Definition 3.8 $F \subset M$ is of first Baire category if it is the union of countably many nowhere dense sets, and of second Baire category otherwise.   □

Remark 3.9 Obviously a subset of a set of first category is of first category, and by Remark 2.5 the union of a countable collection of sets of first category is also of first category. Consequently, if $A = B \cup C$ where A is of second category and B is of first category, then C must be of second category.   □

Lemma 3.10 *Suppose M is of second category and $\cap_{k \in \mathbb{N}} U_k \subset H$, where U_k is open and dense in M for each $k \in \mathbb{N}$. Then H is of second category.*

Proof Since, by hypothesis, $cH \subset \bigcup_{k \in \mathbb{N}} cU_k$ it follows from Lemma 3.4 that cH is of first category, and so by Remark 3.9 H is of second category.   □

Lemma 3.11 *Suppose M is complete and U_k, $k \in \mathbb{N}$, is open and dense in M. Then $\cap_{k \in \mathbb{N}} U_k \neq \emptyset$.*

Proof For $x_1 \in U_1$ let $\epsilon_1 > 0$ be such that $B_{\epsilon_1}(x_1) \subset U_1$. Then, since U_2 is open and dense, there exists ϵ_2 such that

$$B_{\epsilon_2}(x_2) \subset U_2 \text{ and } \overline{B_{\epsilon_2}(x_2)} \subset B_{\epsilon_1}(x_1).$$

Proceeding by induction yields a sequence $\{B_{\epsilon_k}(x_k)\}$ of balls with, for $k \geqslant 2$,

$$B_{\epsilon_k}(x_k) \subset U_k \text{ and } \overline{B_{\epsilon_k}(x_k)} \subset B_{\epsilon_{k-1}}(x_{k-1}), \text{ where } \epsilon_k \to 0 \text{ as } k \to \infty.$$

Hence x_k and $x_{k'}$ are in $B_{\epsilon_K}(x_K)$ for all $k, k' \geqslant K$ and, since $\epsilon_K \to 0$ as $K \to \infty$ and M is complete, it follows that $x_k \to x$ for some $x \in M$. Moreover, since $x_k \in B_{\epsilon_{K+1}}(x_{K+1})$ for all $k > K$ it follows that

$$x \in \overline{B_{\epsilon_{K+1}}(x_{K+1})} \subset B_{\epsilon_K}(x_K) \subset U_K, \text{ for all } K \in \mathbb{N},$$

and the proof is complete.   □

Theorem 3.12 (Baire's Category Theorem [41, p. 139]) *A complete metric space is of second category.*

Proof If M is complete and of first category, $M = \bigcup_{k \in \mathbb{N}} V_k$ where V_k is nowhere dense. Then, by Lemma 3.4, $U_k := c(\overline{V_k})$ is open and dense, and by Lemma 3.11

$$\emptyset \neq \bigcap_{k \in \mathbb{N}} U_k = \bigcap_{k \in \mathbb{N}} c(\overline{V_k}) = c\left(\bigcup_{k \in \mathbb{N}} \overline{V_k}\right) = cM = \emptyset, \text{ which is false.}$$ □

3.3   Paracompactness

Definition 3.13 For an open cover $\mathscr{G}_{\mathscr{A}} = \{G_\alpha : \alpha \in \mathscr{A}\}$ of M,

(i) $\mathscr{G}_{\mathscr{A}}$ is locally finite if every $x \in M$ has a neighbourhood U_x for which the family of indices $\{\alpha \in \mathscr{A} : G_\alpha \cap U_x \neq \emptyset\}$ is finite (note the mapping $\alpha \to G_\alpha$ may not be injective);

(ii) an open cover $\mathscr{H}_{\mathscr{B}} = \{H_\beta : \beta \in \mathscr{B}\}$ of M is a refinement of $\mathscr{G}_{\mathscr{A}}$ if for every $\beta \in \mathscr{B}$ there exists $\alpha \in \mathscr{A}$ such that $H_\beta \subset G_\alpha$;

(iii) M is paracompact if for every open cover $\mathscr{G}_{\mathscr{A}}$ of M there is a locally finite open cover $\mathscr{H}_{\mathscr{B}}$ which is a refinement of $\mathscr{G}_{\mathscr{A}}$. ☐

Theorem 3.14 (Stone [46]) *All metric spaces are paracompact.*

Proof This proof is a reformulation of one due to Mary Ellen Rudin [42].

Let $\mathscr{G}_{\mathscr{A}} = \{G_\alpha : \alpha \in \mathscr{A}\}$ with $\mathscr{A} \neq \emptyset$ be an open cover of M and note that if the family $\mathscr{G}_{\mathscr{A}}$ of sets is finite there is nothing to prove. (Just replace $\mathscr{A}$ by $\mathscr{B} \subset \mathscr{A}$ such that $\alpha \to G_\alpha$ is injective on $\mathscr{B}$ with $\{G_\alpha : \alpha \in \mathscr{B}\} = \{G_\alpha : \alpha \in \mathscr{A}\}$.) So without loss of generality, and to focus the argument, suppose $\mathscr{G}_{\mathscr{A}}$ is an infinite family of distinct sets.

Then, by the well-ordering principle in Chap. 2, let $\preceq$ be a partial ordering such that $(\mathscr{A}, \preceq)$ is a well-ordered set, and write $\alpha' \prec \alpha$ if $\alpha' \preceq \alpha$ and $\alpha' \neq \alpha$.

For all $\alpha \in \mathscr{A}$ let

$$A_\alpha = G_\alpha \setminus \left(\bigcup_{\alpha' \prec \alpha} G_{\alpha'} \right),$$

which may be empty. Clearly $A_\alpha \subset G_\alpha$ and $G_{\alpha'} \cap A_\alpha = \emptyset$ if $\alpha' \prec \alpha$. Moreover, for any $x \in M$, $x \in A_\alpha$ where $\alpha = \min\{\hat{\alpha} \in \mathscr{A} : x \in G_{\hat{\alpha}}\}$, and hence

$$A_\alpha \subset G_\alpha, \quad \cup_{\alpha \in \mathscr{A}} A_\alpha = M, \tag{3.2a}$$

and

$$G_{\alpha'} \cap A_\alpha = \emptyset \text{ or } G_\alpha \cap A_{\alpha'} = \emptyset \text{ if } \alpha \neq \alpha'. \tag{3.2b}$$

Let $\{\mu_n\} \subset (0, \infty)$ be an arbitrary sequence with $\mu_{n-1} > \mu_n \to 0$ as $n \to \infty$ and, with $\mathscr{B} := \mathscr{A} \times \mathbb{N}$, let H_β, $\beta \in \mathscr{B}$, be defined recursively for $n \in \mathbb{N}$ as follows.

Step (i): $n = 1$. When $\beta = (\alpha, 1) \in \mathscr{B}$, $\alpha \in \mathscr{A}$, let

$$A_{\alpha,1} = \{x \in A_\alpha : B_{3\mu_1}(x) \subset G_\alpha\} \text{ and let } H_{(\alpha,1)} = N_{\mu_1}(A_{\alpha,1}), \tag{3.3}$$

where the neighbourhood $N_{\mu_1}(A_{\alpha,1})$ is defined in (3.1a).

Step (ii): $n \geqslant 2$. For $n \geqslant 2$ suppose $H_{\beta'}$ has been defined for all $\beta' = (\alpha', k) \in \mathscr{B}$, $1 \leqslant k < n$ and $\alpha' \in \mathscr{A}$. Then for $\beta = (\alpha, n) \in \mathscr{B}$, $\alpha \in \mathscr{A}$, let

$$A_{\alpha,n} = \left\{ x \in A_\alpha : B_{3\mu_n}(x) \subset G_\alpha \text{ and } x \notin H_{(\alpha',k)} \text{ for all } \alpha' \in \mathscr{A}, \ 1 \leqslant k < n \right\}$$

and, for this $\beta = (\alpha, n) \in \mathscr{B}$, let

$$H_\beta = H_{(\alpha,n)} = N_{\mu_n}(A_{\alpha,n}), \tag{3.4}$$

which is an open subset of G_α. The aim is to show that $\mathscr{H}_{\mathscr{B}} = \{H_\beta : \beta \in \mathscr{B}\}$ is a locally finite open cover of M that is a refinement of $\mathscr{G}_{\mathscr{A}}$.

The next step is to establish two properties of the sets $H_\beta \in \mathscr{H}_{\mathscr{B}}$:

for $n \in \mathbb{N}$ and $\alpha, \alpha' \in \mathscr{A}$ with $\alpha \neq \alpha'$ and $H_{(\alpha,n)} \neq \emptyset \neq H_{(\alpha',n)}$,

$$\inf \left\{ d(x, y) : x \in H_{(\alpha,n)}, \ y \in H_{(\alpha',n)} \right\} \geqslant \mu_n; \tag{3.5}$$

and for $n, m \in \mathbb{N}$ with $n < m$ and any $\alpha, \alpha' \in \mathscr{A}$,

$$\{ x \in H_{(\alpha',m)} : B_{\mu_m}(x) \subset H_{(\alpha,n)} \} = \emptyset. \tag{3.6}$$

Proof of (3.5) By (3.2b) there is no loss in supposing that $G_{\alpha'} \cap A_\alpha = \emptyset$ and by (3.4), with cA defined in (3.1b),

$$N_{2\mu_n}(H_{(\alpha',n)}) \subset N_{3\mu_n}(A_{\alpha',n}) \subset G_{\alpha'} \subset cA_\alpha \subset cA_{\alpha,n},$$

whence $N_{2\mu_n}(H_{(\alpha',n)}) \cap A_{\alpha,n} = \emptyset$. Now (3.5) follows because

$$N_{\mu_n}(H_{(\alpha',n)}) \cap H_{(\alpha,n)} = N_{\mu_n}(H_{(\alpha',n)}) \cap N_{\mu_n}(A_{\alpha,n}) = \emptyset. \qquad \square$$

Proof of (3.6) From the definition of $A_{\alpha',m}$ it is immediate that when $n < m$, $H_{(\alpha,n)} \cap A_{\alpha',m} = \emptyset$, and hence

$$\{ x \in H_{(\alpha',m)} : B_{\mu_m}(x) \subset H_{(\alpha,n)} \}$$
$$\subset \{ x \in N_{\mu_m}(A_{\alpha',m}) : B_{\mu_m}(x) \subset cA_{\alpha',m} \} = \emptyset. \qquad \square$$

Proof that $\mathscr{H}_{\mathscr{B}}$ ***is a Refinement of*** $\mathscr{G}_{\mathscr{A}}$ ***that Covers*** M To see that $\mathscr{H}_{\mathscr{B}}$ covers M let $x \in M$. Then by (3.2a) there exists $\alpha \in A$ such that $x \in A_\alpha \subset G_\alpha$, and there exists $n \geqslant 2$ such that $B_{3\mu_n}(x) \subset G_\alpha$ because G_α is open and $\mu_n \to 0$. Then either $x \in H_{(\alpha,n)}$ and there is nothing more to prove, or $x \notin H_{(\alpha,n)}$ in which case $x \notin A_{\alpha,n}$ because $A_{\alpha,n} \subset H_{(\alpha,n)}$. Therefore, from the definition of $A_{\alpha,n}$, $x \in H_{(\alpha',k)}$ for some $\alpha' \in \mathscr{A}$ and $k \in \{1, \ldots, n-1\}$. Thus $\mathscr{H}_{\mathscr{B}}$ covers M and, since $H_{(\alpha,n)} \subset G_\alpha$ and $H_{(\alpha,n)}$ is open, $\mathscr{H}_{\mathscr{B}}$ and is a refinement of $\mathscr{G}_{\mathscr{A}}$. $\qquad \square$

Proof that $\mathcal{H}_{\mathcal{B}}$ is a Locally Finite Cover of M Let $x \in M$. Then $x \in H_{(\alpha,n)} \subset G_\alpha$ for $(\alpha, n) \in \mathcal{B}$ and so there exists $i_0 > n$ such that $B_{2\mu_{i_0}}(x) \subset H_{(\alpha,n)}$. Hence $B_{\mu_{i_0}+\mu_m}(x) \subset H_{(\alpha,n)}$ for all $m \geq i_0$ since $\{\mu_k\}$ is strictly decreasing. Now suppose $z \in B_{\mu_{i_0}}(x)$. Then $B_{\mu_m}(z) \subset H_{(\alpha,n)}$ for all $m \geq i_0$, and it follows by (3.6) that $z \notin H_{(\alpha',m)}$ for all $\alpha' \in \mathcal{A}$. Therefore, $B_{\mu_{i_0}}(x) \cap H_{(\alpha',m)} = \emptyset$ for all $m \geq i_0$ and all $\alpha' \in \mathcal{A}$. However by (3.5), for each $m < i_0$, there is at most one $\alpha' \in \mathcal{A}$ such that $B_{\frac{1}{2}\mu_{i_0}}(x) \cap H_{(\alpha',m)} \neq \emptyset$. Hence the set of indices in $\mathcal{B}$ such that the corresponding elements in $\mathcal{H}_{\mathcal{B}}$ intersects $B_{\frac{1}{2}\mu_{i_0}}(x)$ is at most finite. This completes the proof of Theorem 3.14. $\qquad\square$

3.4 Metric Subspaces

For $\emptyset \neq A \subset M$ the metric subspace defined by restricting d to $A \times A$ is denoted by (A, d_A). Then $Y \subset A$ is open in (A, d_A) if and only $Y = A \cap G$ for an open G in M, and Y is closed in (A, d_A) if and only $Y = A \cap F$ for a closed F in M. Such sets are said to be open or closed in A, or just relatively open or closed when the set A is understood.

When $X \subset A \subset M$, let $\overline{X}^A$, X°^A} and $\partial_A X$ denote the closure, interior and boundary of X in (A, d_A). Then

$$\overline{X}^A = A \cap \overline{X}, \tag{3.7a}$$

and hence by (3.1c),

$$X^{\circ^A} = A \setminus \overline{(A \setminus X)}^A = A \setminus \left(\overline{(A \setminus X)} \cap A \right)$$
$$= A \setminus \overline{(A \setminus X)} = A \cap (X \cup cA)^\circ \tag{3.7b}$$
$$\supset A \cap X^\circ = X^\circ, \text{ for all } X \subset A.$$

Moreover, when X is open in M, then $X = A \cap X$ is also open in A, and thus $X^{\circ^A} = X = X^\circ$. When A is open, for all $x \in X^{\circ^A}$ there exists $\epsilon > 0$ such that $B_\epsilon(x) \cap A \subset X$, which implies $B_\epsilon(x) \subset A \subset X$ if ϵ is small enough and thus $X^{\circ^A} \subset X^\circ$. In summary,

$$X^{\circ^A} \supset X^\circ \text{ for all } X \subset A, \text{ and } X^\circ = X^{\circ^A} \text{ if } A \text{ or } X \text{ is open in } M, \tag{3.7c}$$

and consequently

$$\partial_A X \subset A \cap \partial X, \text{ and equality holds when } X \text{ or } A \text{ is open.} \tag{3.7d}$$

As a consequence of the above definitions of open and closed sets in A, a set $X \subset A$ is open or closed in A exactly when $X = X^{\circ^A}$ or $X = \overline{X}^A$. By definition, non-empty subsets A of M have properties such as in Definition 3.2 if the metric space (A, d_A) has those properties. Also, A is said to be relatively compact in M if $\overline{A}$ is compact in M. By definition the empty set $\emptyset$ is both open and closed in A, and compact.

3.5 Connected Sets

Definition 3.15 A metric space M is separated if $M = G_1 \cup G_2$, where $G_i, i = 1, 2$, are non-empty, disjoint open sets. Since, for $i \neq j$, $G_i = M \setminus G_j$, the sets G_i, $i = 1, 2$, are also closed, and the pair $\{G_1, G_2\}$ is called a separation of M. More generally, $A \subset M$ is separated if $A \neq \emptyset$ and (A, d_A) is separated. In other words, for some $G_i, i = 1, 2$, open in M,

$$A \subset G_1 \cup G_2, \quad G_1 \cap A \neq \emptyset \neq G_2 \cap A \text{ and } G_1 \cap G_2 \cap A = \emptyset.$$

Then $\{A \cap G_1, A \cap G_2\}$ is a separation of (A, d_A). By definition, $\emptyset$ is not separated. $\square$

Note the difference between metric spaces being separable (Definition 3.5) and separated (Definition 3.15).

Lemma 3.16 *A set $A \subset M$ is separated if and only if for a pair of non-empty subsets B and C of M,*

$$A = B \cup C \text{ and } B \cap \overline{C} = \overline{B} \cap C = \emptyset, \tag{3.8}$$

in which case $\{B, C\}$ is a separation of A.

Proof Suppose (3.8) holds. Then

$$B = (B \cup C) \cap \overline{B} = A \cap \overline{B} \text{ and } C = (B \cup C) \cap \overline{C} = A \cap \overline{C},$$

and it follows that B and C are closed in A, disjoint by (3.8), and non-empty by hypothesis. Hence the metric subspace (A, d_A) is separated and $\{B, C\}$ is a separation of A.

Conversely if A is separated, $A = A_1 \cup A_2$ where A_1, A_2 are non-empty disjoint sets which are closed in A, and so

$$\emptyset = A_1 \cap A_2 = (A \cap \overline{A_1}) \cap A_2 = A_2 \cap \overline{A_1},$$

and similarly $\overline{A_2} \cap A_1 = \emptyset$. Thus $B = A_1$ and $C = A_2$ satisfies (3.8) if A is separated. $\square$

Lemma 3.17 *If $A \subset B \subset \overline{A}$ where B is separated, then A is separated.*

Proof Since B is separated there exist open sets G_i, $i = 1, 2$, in M with

$$G_i \cap B \neq \emptyset, \; i = 1, 2, \quad B \subset G_1 \cup G_2 \; \text{and} \; G_1 \cap G_2 \cap B = \emptyset.$$

Then $A \subset G_1 \cup G_2$, $G_1 \cap G_2 \cap A = \emptyset$, and $G_i \cap A \neq \emptyset$, $i = 1, 2$, because G_i is open, $G_i \cap B \neq \emptyset$ and $B \subset \overline{A}$. Hence $\{G_1 \cap A, G_2 \cap A\}$ separates A. □

Definition 3.18 A metric space M is connected if it is not separated, and a connected metric space is non-degenerate if it has more than one point. A non-empty set A in M is connected if and only if (A, d_A) is connected, and $\emptyset$ is connected. □

Remark 3.19 A metric space M is connected if and only if no proper subset of M is both open and closed. If subsets A and B of M are connected and $A \cap B \neq \emptyset$, then $A \cup B$ is connected. More generally, for any non-empty set $\mathscr{A}$ of indices, if $\cap_{\alpha \in \mathscr{A}} C_\alpha \neq \emptyset$ and C_α is a connected subset of M for all $\alpha \in \mathscr{A}$, then $\cup_{\alpha \in \mathscr{A}} C_\alpha$ is connected. A singleton $\{p\}$ is always connected, and the empty set $\emptyset$, which is open and closed, is connected. □

Lemma 3.20 *If (M_i, d_i), $i = 1, 2$, are connected metric spaces, then the product space $M_1 \times M_2$, see (3.1d), is connected.*

Proof Let $(\hat{x}, \tilde{y})$ and $(\tilde{x}, \hat{y})$ be arbitrary points of $M_1 \times M_2$ and note that the functions $f_i : M_i \to M_1 \times M_2, i = 1, 2$, defined by

$$f_1(y) = (\hat{x}, y), \; y \in M_2, \quad \text{and} \quad f_2(x) = (x, \hat{y}), \; x \in M_1,$$

are continuous and hence their ranges, $\{\hat{x}\} \times M_2$ and $M_1 \times \{\hat{y}\}$, are connected in $M_1 \times M_2$. Hence, since $(\hat{x}, \hat{y})$ is in both, it follows from Remark 3.19 that their union $(\{\hat{x}\} \times M_2) \cup (M_1 \times \{\hat{y}\})$ is a connected set in $M_1 \times M_2$.

But since $(\hat{x}, \tilde{y})$ and $(\tilde{x}, \hat{y})$ also belong to their union, this shows that for any pair of points in $M_1 \times M_2$ there is a connected set which contains them both, and it follows easily that $M_1 \times M_2$ is connected. □

Definition 3.21 A non-empty subset A of M is a continuum if (A, d_A) is a compact, connected metric space, and a generalised continuum if (A, d_A) is a connected and locally compact metric space. As in Definition 3.18, a continuum is non-degenerate if it has more than one point. □

Remark It is worth noting that a generalised continuum A in M is not necessarily closed in M. For example, $A = (0, 1)$ is a generalised continuum in $M = [0, 1]$ which is not closed in M. □

Remark 3.22 To see that a connected set $E \subset \mathbb{R}$ is an interval, suppose it is not. Then there are points a, $c \in E$ and $b \notin E$ with $a < b < c$ and $E = (E \cap (-\infty, b)) \cup (E \cap (b, \infty))$, which separates E.

To see that the interval $[a, b]$ is connected, suppose it is not. Then there exist disjoint, non-empty, closed sets F_1 and F_2 such that $[a, b] = F_1 \cup F_2$ and by compactness there exists $f_1 \in F_1$ and $f_2 \in F_2$ such that

$$|f_1 - f_2| = \text{dist}\,(F_1, F_2) := \inf\{|x - y| : x \in F_1, y \in F_2\} > 0.$$

But then $z = \frac{1}{2}(f_1 + f_2) \in [a, b]$, say $z \in F_1$. Then $|z - f_2| = \frac{1}{2}|f_1 - f_2| < |f_1 - f_2|$, a contradiction. Therefore, since every non-empty interval can be written as an union of bounded closed intervals with a common point (see Remark 3.19), every non-empty interval is connected. $\qquad\square$

Lemma 3.23 *If $f : M \to N$ is continuous where M and N are metric spaces and M is connected, then $f(M)$ is connected.*

Proof If this is false there exist open sets G_1, G_2 in N such that

$$f(M) \cap G_1 \neq \emptyset \neq f(M) \cap G_2, \quad f(M) \cap G_1 \cap G_2 = \emptyset \text{ and } f(M) \subset G_1 \cup G_2.$$

Then $\widetilde{G}_i = f^{-1}(G_i)$, $i = 1, 2$, are disjoint, non-empty, open sets that cover M. Since M is connected this is false, and hence $f(M)$ is connected. $\qquad\square$

Corollary 3.24 *A non-degenerate, connected metric space is uncountable.*

Proof For distinct points y, z in a non-degenerate, connected metric space M, define a continuous function $f : M \to \mathbb{R}$ by $f(x) = d(x, y)$, $x \in M$. Then since $f(y) = 0 < f(z)$ and $f(M)$ is connected by Lemma 3.23, it follows by Remark 3.22 that $[0, f(z)] \subset f(M)$. Therefore $f(M)$ is uncountable because $[0, f(z)]$ is uncountable by Remark 2.5, and so M is uncountable. $\qquad\square$

Lemma 3.25 *If A is connected and $A \subset B \subset \overline{A}$, then B is connected. In particular, $\overline{A}$ is connected when A is connected.*

Proof This is immediate from Lemma 3.17 and Definition 3.18. $\qquad\square$

Lemma 3.26 *Suppose M is connected, $U \subset M$ and $\overline{U} = W_1 \cup W_2$, where W_1 and W_2 are non-empty, closed and $W_1 \cap W_2 = \emptyset$. Then $\overline{U} \neq M$, $\partial U \cap W_1 \neq \emptyset$ and $\partial U \cap W_2 \neq \emptyset$. In other words, ∂U is not connected.*

Proof Since M is connected, $\overline{U} \neq M$. Let $V = M \setminus (U^\circ)$ and note that

$$M = V \cup (U^\circ) \subset V \cup \overline{U} = (V \cup W_1) \cup W_2.$$

Now note that $(V \cup W_1) \cap W_2 \neq \emptyset$ since either $V \cup W_1 = M$, or if not because M is connected and V and W_i, $i = 1, 2$, are closed. Therefore

$$\emptyset \neq (V \cup W_1) \cap W_2 = V \cap W_2 = W_2 \setminus (U^\circ) = (W_2 \cap \overline{U}) \setminus (U^\circ) = W_2 \cap \partial U.$$

To see that $\partial U \cap W_1 \neq \emptyset$, exchange the roles of W_1 and W_2. $\qquad\square$

Corollary 3.27 *For $U \subset M$, if ∂U and M are connected, then both $\overline{U}$ and $\overline{cU}$ are connected.*

Proof If $\overline{U}$ is not connected, $U \neq \emptyset$ and there exist W_1 and W_2 which satisfies the hypotheses of the lemma. Hence ∂U is not connected which is a contradiction. It follows that $\overline{cU}$ is connected because $\partial U = \partial(cU)$. ☐

Theorem 3.28 ([43, Thm. 2, p. 4]) *A connected, locally compact metric space is σ-compact, and hence separable.*

Proof If M is locally compact, it has an open cover $\mathscr{G}_{\mathscr{A}} = \{G_\alpha : \alpha \in \mathscr{A}\}$ for which each $\overline{G}_\alpha$ is compact and hence, by Theorem 3.14, there is an open cover $\mathscr{H}_{\mathscr{B}} = \{H_\beta : \beta \in \mathscr{B}\}$ of M which is a locally finite refinement of $\mathscr{G}_{\mathscr{A}}$ for which all the sets $\overline{H}_\beta$, $\beta \in \mathscr{B}$, are compact.

Therefore, for any $\beta_0 \in \mathscr{B}$ such that $H_{\beta_0} \neq \emptyset$, there is a finite set $\{H_{\beta_1}, \cdots, H_{\beta_{n_0}}\} \subset \mathscr{H}_{\mathscr{B}}$ with

$$\overline{H}_{\beta_0} \subset H_{\beta_0} \cup H_{\beta_1} \cup \cdots \cup H_{\beta_{n_0}}$$

because $\{H_\beta \cap \overline{H}_{\beta_0} : \beta \in \mathscr{B}\}$ is an open cover of the compact metric subspace $\overline{H}_{\beta_0}$. Then by the same argument there is a finite set $H_{\beta_{n_0+1}}, \cdots, H_{\beta_{n_1}}$ such that

$$\overline{H}_{\beta_0} \cup \cdots \cup \overline{H}_{\beta_{n_0}} \subset H_{\beta_0} \cup \cdots \cup H_{\beta_{n_0}} \cup H_{\beta_{n_0+1}} \cup \cdots \cup H_{\beta_{n_1}},$$

and by induction there is a sequence $\{\beta_k\}$ of not necessarily distinct elements of $\mathscr{B}$ such that

$$H := \bigcup_{k=0}^{\infty} H_{\beta_k} \subset \bigcup_{k=0}^{\infty} \overline{H}_{\beta_k} \subset \bigcup_{k=0}^{\infty} H_{\beta_k} = H.$$

Then H is open because $\mathscr{H}_{\mathscr{B}}$ is an open cover of M. To see that H is closed, let $x_i \to x$ where $x_i \in H_{\beta_{k_i}} \subset H$, and $x \in H_\beta$ for some $\beta \in \mathscr{B}$. Since $\mathscr{H}_{\mathscr{B}}$ is a locally finite open cover of M, there exists $\delta > 0$ such that $B_\delta(x) \subset H_\beta$ and only a finite subset of the sets $\{H_{\beta_{k_i}} : i \in \mathbb{N}\}$ intersect $B_\delta(x)$. Hence, since $x_i \to x \in H_\beta$, the sequence $\{x_i\}$ must lie in a finite union of sets $H_{\beta^1} \cup \cdots \cup H_{\beta^m}$, where $\{\beta^j, 1 \leqslant j \leqslant m\} \subset \{\beta_{k_i} : i \in \mathbb{N}\}$. This shows that $x \in \overline{H}_{\beta^1} \cup \cdots \cup \overline{H}_{\beta^m} \subset H$ and hence H is closed.

Since M is connected, it follows that $M = H$ where H is the union of a countable family of compact sets. Therefore M is σ-compact and hence separable. This completes the proof. ☐

3.6 Simple Chains

Definition 3.29 A finite family $\mathscr{G} = \{G_1, \cdots, G_n\}$ of subsets of a set A is a simple chain in A if

$$G_i \cap G_j \neq \emptyset \quad \text{if and only if} \quad |i - j| \leqslant 1, \tag{3.9a}$$

and for a simple chain

$$[\mathscr{G}] := \bigcup_{i=1}^{n} G_i \subset A. \tag{3.9b}$$

The sets G_i are the links of $\mathscr{G}$ and a simple chain is said to be open in A if all its links are open in A. The links G_i and G_j are adjacent when $|i - j| = 1$ and a link with only one adjacent link is called an end link. (Note that an end link can be a subset of its adjacent link.) A simple chain in A is said to join (not necessarily distinct) points x and y in A if

$$x \in G_i \text{ if and only if } i = 1 \text{ and } y \in G_j \text{ if and only if } j = n. \tag{3.10}$$

Note that if $x = y$, then $n = 1$ and $x \in G_1$. $\qquad\qquad\qquad\qquad\qquad\qquad\square$

Remark 3.30 By Remark 3.19, $[\mathscr{G}]$ is connected if all the links of $\mathscr{G}$ are connected. Moreover, $[\mathscr{G}]$ is open, closed or compact in A if all its links are open, closed or compact in A. $\qquad\qquad\qquad\qquad\qquad\qquad\qquad\qquad\qquad\qquad\qquad\qquad\square$

Definition 3.31 For simple chains $\mathscr{U} = \{U_1, \cdots, U_m\}$ and $\mathscr{V} = \{V_1, \cdots, V_n\}$, $\mathscr{U}$ is a sub-chain of $\mathscr{V}$ if each link of $\mathscr{U}$ is a link of $\mathscr{V}$, which is written $\mathscr{U} = \mathscr{V}(i, j)$ if $U_1 = V_i$ and $U_n = V_j$. (Note that $i > j$, $i < j$ and $i = j$ are possibilities.) $\qquad\square$

Lemma 3.32 *Suppose* $\mathscr{G} = \{G_1, \cdots, G_n\}$ *is a simple chain the links of which are either all open or all closed, and for* $H \subset [\mathscr{G}]$,

$$H \cap G_k = \emptyset \text{ for some } k, \ 1 < k < n, \text{ and } H \cap G_i \neq \emptyset \neq H \cap G_j,$$

where $1 \leqslant i < k < j \leqslant n$. *Then H is not connected.*

Proof Since, by Definition 3.29, $\left(\cup_{i=1}^{k-1} G_i\right) \cap \left(\cup_{j=k+1}^{n} G_j\right) = \emptyset$ and, by hypothesis, $H \subset \left(\cup_{i=1}^{k-1} G_i\right) \cup \left(\cup_{j=k+1}^{n} G_j\right)$ where $\mathscr{G}$ is a simple chain with links that are all either open or closed, this yields a separation (Definition 3.15) of H, and hence H is not connected. $\qquad\qquad\qquad\qquad\qquad\qquad\qquad\qquad\qquad\qquad\qquad\qquad\square$

Theorem 3.33 *If* $\mathscr{G}_{\mathscr{A}} = \{G_\alpha : \alpha \in \mathscr{A}\}$ *is an open cover of a connected metric space M, and $x, y \in M$, there is a simple chain with links in $\mathscr{G}_{\mathscr{A}}$ that joins x to y.*

Proof Fix $x \in M$ and let B be the set of $y \in M$ for which there is a simple chain with links in $\mathscr{G}_{\mathscr{A}}$ joining x to y. The goal is to show that B is non-empty, open and closed, since that implies $B = M$ because M is connected.

To see that B is non-empty and open, first note that $x \in G_\alpha$ for some $\alpha \in \mathscr{A}$, since $\mathscr{G}_{\mathscr{A}}$ covers M, which implies that $G_\alpha \subset B$ because $\{G_\alpha\}$ is a chain with one link that joins x to any element of G_α. Then since $\mathscr{G}_{\mathscr{A}}$ is open,

$$U_x := \bigcup_{\substack{\alpha \in \mathscr{A} \\ x \in G_\alpha}} G_\alpha \quad \text{is an open subset of } B.$$

Now if $y \in B \setminus U_x$ there is a simple chain $\{G_{\alpha_1}, \cdots, G_{\alpha_n}\} \subset \mathscr{G}_{\mathscr{A}}$ joining x to y with $n \geqslant 2$. Then $y \in G_{\alpha_n} \setminus G_{\alpha_{n-1}}$ by (3.10), and the same chain shows that (3.10) holds for any point of $G_{\alpha_n} \setminus G_{\alpha_{n-1}}$, and hence $G_{\alpha_n} \setminus G_{\alpha_{n-1}} \subset B$. If $n = 2$, then $G_{\alpha_1} \subset U_x \subset B$ and hence $G_{\alpha_2} = G_{\alpha_n} \subset B$. If $n \geqslant 3$, it follows by (3.9a) that $(G_{\alpha_n} \cap G_{\alpha_{n-1}}) \cap G_{n-2} = \emptyset$, and hence $G_{\alpha_n} \cap G_{\alpha_{n-1}} \subset B$ because $\{G_{\alpha_1}, \cdots, G_{\alpha_{n-1}}\}$ is a simple chain joining x to any point of $G_{\alpha_n} \cap G_{\alpha_{n-1}}$.

Thus, for $n \geqslant 2$, G_{α_n} is a subset of B and, since G_{α_n} is open, every $y \in B \setminus U_x$ is an interior point of B. It follows that $B = U_x \cup (B \setminus U_x)$ is open, as required.

To show that B is closed, let $z \in \overline{B}$. Then $z \in G_\alpha$ for some $\alpha \in \mathscr{A}$, and there exists $y \in B \cap G_\alpha$ since G_α is open.

If $x \in G_\alpha$, then $\{G_\alpha\}$ is a simple chain joining x to z and thus $z \in B$. If $x \notin G_\alpha$, let $\{G_{\alpha_1}, \cdots, G_{\alpha_n}\} \subset \mathscr{G}_{\mathscr{A}}$ be a simple chain joining x to y. If $z \in G_{\alpha_i}$ for some $i = 1, \cdots, n$, let k be the smallest such i. Then $\{G_{\alpha_1}, \cdots, G_{\alpha_k}\}$ is a simple chain joining x to z, and $z \in B$.

If $z \notin G_{\alpha_i}$ for all $i = 1, \cdots, n$, let k be the smallest $i \in \{1, \cdots, n\}$ such that $G_\alpha \cap G_{\alpha_i} \neq \emptyset$. Then $\{G_{\alpha_1}, \cdots, G_{\alpha_k}, G_\alpha\}$ is a simple chain joining x to z, and again $z \in B$. This shows that B is closed.

Since $B \neq \emptyset$ is open and closed and M is connected, $M = B$, as required. □

Corollary 3.34 *When $A \subset M$ is connected, for $\epsilon > 0$ and $x, y \in A$, there exists (a) $\{x_1, \cdots, x_n\} \subset A$ and $\{\epsilon_1, \cdots, \epsilon_n\} \subset (0, \epsilon)$ with $x = x_1$, $y = x_n$, $x \in B_{\epsilon_i}(x_i)$ only if $i = 1$, $y \in B_{\epsilon_i}(x_i)$ only if $i = n$, $1 \leqslant i \leqslant n$, and*

$$A \cap B_{\epsilon_i}(x_i) \cap B_{\epsilon_j}(x_j) \neq \emptyset \text{ if and only if } |i - j| \leq 1, \quad 1 \leq i, j \leq n. \tag{3.11}$$

(b) When A is open, (a) holds with $B_{\epsilon_i}(x_i) \subset A$, $1 \leqslant i \leqslant n$.

Proof Since A is connected and

$$\mathscr{G} = \{A \cap B_\delta(z) : \delta \in (0, \epsilon), z \in \{x, y\}\} \bigcup$$
$$\{A \cap B_\delta(z) : \delta \in (0, \epsilon), \{x, y\} \cap B_\delta(z) = \emptyset, z \in A\}$$

is an open cover of A in (A, d_A), (a) follows from Theorem 3.33. When A is open and connected in M, (b) follows likewise since

$$\mathscr{G} = \{B_\delta(z) \subset A : \delta \in (0, \epsilon), z \in \{x, y\}\} \bigcup$$
$$\{B_\delta(z) \subset A : \delta \in (0, \epsilon), \{x, y\} \cap B_\delta(z) = \emptyset, z \in A\}$$

is an open cover of A. □

Definition 3.35 For any $A \subset M$ and $x, y \in A$, an ϵ-chain joining x to y in A is a finite set $\{x_i^\epsilon : 1 \leqslant i \leqslant n_\epsilon\} \subset A$ such that

$$x_1^\epsilon = x, \quad x_{n_\epsilon}^\epsilon = y \quad \text{and} \quad d(x_j^\epsilon, x_{j+1}^\epsilon) < \epsilon, \quad 1 \leqslant j \leqslant n_\epsilon - 1,$$

A is ϵ-chained if every pair $x, y \in A$ is connected by an ϵ-chain, and A is chained if it is ϵ-chained for all $\epsilon > 0$. ☐

Lemma 3.36 *Connected sets are chained.*

Proof This is immediate by Corollary 3.34. ☐

Lemma 3.37 *For $\emptyset \neq A \subset M$ and $\epsilon > 0$, the set A_ϵ of points $x \in M$ which can be joined to some point of A by an ϵ-chain in M, is open and closed.*

Proof If $x \in A_\epsilon$, a point $a \in A$ is joined to x by an ϵ-chain, and so every point of $B_\epsilon(x)$ is joined to a by an ϵ-chain. Hence A_ϵ is open.

To show that A_ϵ is closed, suppose that $y \in \overline{A_\epsilon}$. Then there exists $x \in A_\epsilon$ with $d(x, y) < \epsilon$, which implies that $y \in A_\epsilon$. Hence A_ϵ is closed. ☐

3.7 Sequences of Sets

Definition 3.38 For a sequence $\{A_j\}_{j \in \mathbb{N}}$ of non-empty sets in M

$$\limsup A_j = \left\{ x \in M : \text{for all } \delta > 0, \{j \in \mathbb{N} : B_\delta(x) \cap A_j \neq \emptyset\} \text{ is infinite} \right\}$$
$$= \left\{ x \in M : \text{for all } \delta > 0, \{j \in \mathbb{N} : \text{dist}(x, A_j) < \delta\} \text{ is infinite} \right\},$$

and

$$\liminf A_j = \left\{ x \in M : \text{for all } \delta > 0, \{j \in \mathbb{N} : B_\delta(x) \cap A_j = \emptyset\} \text{ is finite} \right\}$$
$$= \left\{ x \in M : \text{for all } \delta > 0 \{j \in \mathbb{N} : \text{dist}(x, A_j) > \delta\} \text{ is finite} \right\},$$

or equivalently,

$$\limsup A_j = \bigcap_{\delta > 0} \bigcap_{k \geqslant 1} \bigcup_{j \geqslant k} N_\delta(A_j) \quad \text{and} \quad \liminf A_j = \bigcap_{\delta > 0} \bigcup_{k \geqslant 1} \bigcap_{j \geqslant k} N_\delta(A_j).$$

☐

Since $\text{dist}(x, A_j) = \text{dist}(x, \overline{A_j})$ it is easily checked that

(i) $\liminf A_j \subset \limsup A_j$;
(ii) $\limsup A_j = \limsup \overline{A_j}$, $\quad \liminf A_j = \liminf \overline{A_j}$;
(iii) if $A_j \subset B_j$, $j \in \mathbb{N}$, then $\liminf A_j \subset \liminf B_j$, $\limsup A_j \subset \limsup B_j$;

(iv) for any subsequence $\{A_{k_j}\}$ of $\{A_j\}$,
$$\liminf A_j \subset \liminf A_{k_j} \subset \limsup A_{k_j} \subset \limsup A_j;$$
(v) $\liminf A_j$ and $\limsup A_j$ are closed in M because, for all $\delta > 0$,

$$\overline{\bigcup_{j \geqslant k} N_\delta(A_j)} \subset \bigcup_{j \geqslant k} N_{2\delta}(A_j), \quad \overline{\bigcup_{k \geqslant 1} \bigcap_{j \geqslant k} N_\delta(A_j)} \subset \bigcup_{k \geqslant 1} \bigcap_{j \geqslant k} N_{2\delta}(A_j),$$

and for any family $\{Y_\gamma\}$ of sets, $\overline{\bigcap_\gamma Y_\gamma} \subset \bigcap_\gamma \overline{Y_\gamma}$;

(vi) $\left(\bigcup_j \overline{A}_j\right) \cup (\limsup A_j)$ is closed in M;

(vii) $\limsup A_j \neq \emptyset$ when $A = \bigcup_{j \in \mathbb{N}} A_j$ is relatively compact;

(viii) if a sequence $\{A_j\}$ is nested (meaning $A_{j+1} \subset A_j$ for all $j \in \mathbb{N}$), $\left(\bigcap_j A_j\right) \subset \liminf A_j = \limsup A_j$.

Lemma 3.39 *Let $S = \limsup A_j$, where $\{A_j\}$ is a sequence of non-empty sets for which $A = \bigcup_{j \in \mathbb{N}} A_j$ is relatively compact. Then for all $\epsilon > 0$ there exists $J \in \mathbb{N}$ such that $A_j \subset N_\epsilon(S)$ for all $j > J$.*

Proof Suppose this is false. Then for some $\epsilon > 0$ and an increasing sequences $\{j_k\}$ of distinct integers there exists $x_k \in A_{j_k} \setminus N_\epsilon(S)$. Since $\overline{A}$ is compact there is no loss in assuming that $x_k \to x$ as $k \to \infty$, and $x \in S$ because the sequence $\{j_k\}$ is unbounded. But $x \notin N_\epsilon(S)$ because $x_k \notin N_\epsilon(S)$ for all $k \in \mathbb{N}$. This contradiction proves the lemma.   □

In the next result it is not assumed that the sets A_j are connected.

Theorem 3.40 *Let $\{A_j\}$ be a sequence of non-empty sets such that $A = \bigcup_{j \in \mathbb{N}} A_j$ is relatively compact and for all pairs x and $y \in A_j$, $x \neq y$, suppose there is an ϵ_j-chain joining x to y in A_j with $\epsilon_j \to 0$ as $j \to \infty$. Then $\limsup A_j$, which is non-empty by (vii), is connected if $\liminf A_j \neq \emptyset$.*

Proof Since A is relatively compact, $S := \limsup A_j$ is non-empty by (vii) above, and compact because $S \subset \overline{A}$ is closed by (v).

Suppose $I := \liminf A_j \neq \emptyset$ and S is not connected. Then

$$S = F_1 \cup F_2 \text{ where } F_i \neq \emptyset, \ i = 1, 2, \text{ are compact sets with } F_1 \cap F_2 = \emptyset,$$

and since $I \subset S$, there is no loss in assuming that $I \cap F_1 \neq \emptyset$. Now let $d = \text{dist}\,(F_1, F_2) > 0$ and let $\delta = d/3$. Since $\epsilon_j \to 0$, by Lemma 3.39 there exists J such that

$$\epsilon_j < d/4 \text{ and } A_j \subset N_\delta(S) = N_\delta(F_1) \cup N_\delta(F_2), \text{ for all } j > J.$$

Choose $x \in I \cap F_1$. Then, if J is chosen large enough,

$$\text{for all } j > J \ \emptyset \neq B_\delta(x) \cap A_j \subset N_\delta(F_1) \cap A_j.$$

Therefore, since $\emptyset \neq F_2 \subset S$, there exists $j^* > J$ with $\epsilon_{j^*} < d/4$

$$N_\delta(F_1) \cap A_{j^*} \neq \emptyset, \ N_\delta(F_2) \cap A_{j^*} \neq \emptyset \text{ and } A_{j^*} \subset N_\delta(F_1) \cup N_\delta(F_2). \qquad (3.12)$$

But (3.12) is impossible because the hypothesis states that any two points of A_{j^*} are joined by an ϵ_{j^*}-chain where

$$\epsilon_{j^*} < d - \frac{2}{3}d = \text{dist}\,(F_1, F_2) - 2\delta \leqslant \text{dist}\,(N_\delta(F_1), N_\delta(F_2)).$$

This contradiction completes the proof. $\qquad\qquad\qquad\qquad\qquad\qquad\qquad\qquad$ □

Corollary 3.41 *Suppose, for a sequence $\{A_j\}$ of non-empty connected sets, that $A = \bigcup_{j \in \mathbb{N}} A_j$ is relatively compact and $\liminf A_j \neq \emptyset$. Then $\limsup A_j$ is connected.*

Proof This is immediate from Lemma 3.36 and Theorem 3.40. $\qquad\qquad\qquad\qquad$ □

3.8 Convergent Sequences of Sets

Definition 3.42 A sequence $\{A_j\}$ of non-empty sets in M is convergent to A if $A = \limsup A_j = \liminf A_j$, written $A_j \to A$ or $\lim A_j = A$. $\qquad\qquad\qquad$ □

Theorem 3.43 *When $\{K_j\}$ is a nested sequence of continua (Definition 3.21), $\lim K_j = \bigcap_{j \in \mathbb{N}} K_j \neq \emptyset$ is a continuum.*

Proof Since $K_j \neq \emptyset$ is compact and $K_{j+1} \subset K_j$ for all j, the intersection $K = \bigcap_{j \in \mathbb{N}} K_j$ is non-empty and $K \subset \liminf K_j = \limsup K_j$ by (viii) in Sect. 3.7. Therefore, by Corollary 3.41, $\limsup K_j$ is a continuum. Now if $x \in \limsup K_j$, there exists a sequence $\{j_k\} \subset \mathbb{N}$ and $x_{j_k} \in K_{j_k}$ such that $j_k \to +\infty$ and $x_{j_k} \to x$, and hence $x \in K_j$ for all j. Therefore, $K \subset \liminf K_j = \limsup K_j \subset K$, and hence $K = \lim K_j$ is a continuum. $\qquad\qquad\qquad\qquad\qquad\qquad$ □

Theorem 3.44 *Any sequence $\{A_j\}$ of non-empty subsets of a separable metric space M has a convergent subsequence, and if $\limsup A_j \neq \emptyset$ the convergent subsequence can be chosen so that its limit is non-empty.*

Proof If $\limsup A_j = \emptyset$ then by (i), Sect. 3.7, $\lim A_j = \emptyset$. So assuming that $\limsup A_j \neq \emptyset$, let $x \in \limsup A_j$ and let $\{j_k\}$ be an increasing sequence such that $B_{1/k}(x) \cap A_{j_k} \neq \emptyset$, $k \in \mathbb{N}$. Then replacing, if necessary, $\{A_j\}$ with its subsequence $\{A_{j_k}\}$, there is no loss in assuming that $x \in \liminf A_j$.

Since M is separable (Definition 3.5 and Remark 3.7) it has a countable base $\mathscr{W} = \{W_n : n \in \mathbb{N}\}$ which will be used to define a convergent subsequence of $\{A_j\}$ as the diagonal $\{A_{j,j}\}$ of a sequence $\{A_{1,j}\}, \{A_{2,j}\}, \{A_{3,j}\}, \ldots$ of subsequences of $\{A_j\}$ as follows.

Let $\{A_{1,j}\} = \{A_j\}$ and define $\{A_{n+1,j}\}$ for $n \in \mathbb{N}$ recursively as follows.

(α) If the $\limsup$ of every subsequence of $\{A_{n,j}\}$ intersects W_n let

$$\{A_{n+1,j}\} = \{A_{n,j}\}.$$

(β) Otherwise let

$\{A_{n+1,j}\}$ be a subsequence of $\{A_{n,j}\}$ such that $\limsup A_{n+1,j} \cap W_n = \emptyset$.

This yields an array

$$
\begin{array}{cccccc}
A_{1,1} & A_{1,2} & A_{1,3} & \cdots & A_{1,m+1} & \cdots \\
A_{2,1} & A_{2,2} & A_{2,3} & \cdots & A_{2,m+1} & \cdots \\
\vdots & \vdots & \ddots & \vdots & \vdots & \vdots \\
A_{m,1} & A_{m,2} & \cdots & A_{m,m} & A_{m,m+1} & \cdots \\
\vdots & \vdots & \vdots & \vdots & \vdots & \vdots
\end{array}
$$

in which $\{A_{j,j} : j \geqslant m\}$ is a subsequence of $\{A_{m,j} : j \geqslant m\}$ for all $m \in \mathbb{N}$.

Now suppose $a \notin \liminf A_{j,j}$. Then there is $\delta > 0$ and a subsequence, $\{A_{j_\ell,j_\ell}\}$ say, of $\{A_{j,j}\}$, such that $A_{j_\ell,j_\ell} \cap B_\delta(a) = \emptyset$ for all ℓ. Then since $\mathscr{W}$ is a base for M, $a \in W_m \subset B_{\delta/2}(a)$ for some m and, from the definition of $\limsup$,

$$\limsup A_{j_\ell,j_\ell} \cap W_m = \emptyset. \tag{3.13}$$

Now, since $\{A_{j_\ell,j_\ell} : \ell \geqslant m\}$ is a subsequence of $\{A_{m,j} : j \geqslant m\}$, it follows from ($\beta$) that $W_m \cap \limsup A_{m+1,j} = \emptyset$ and, since $\{A_{j,j} : j \geqslant m+1\}$ is a subsequence of $\{A_{m+1,j} : j \geqslant m+1\}$, it follows that

$$W_m \cap \limsup A_{j,j} = \emptyset \tag{3.14}$$

by (iv) in Sect. 3.7.

Hence $a \notin \liminf A_{j,j}$ implies $a \notin \limsup A_{j,j}$ and so, by (i) in Sect. 3.7, $\liminf A_{j,j} = \limsup A_{j,j}$. Thus $\{A_{j,j}\}$ is convergent and $\lim A_{j,j} \neq \emptyset$ because, by (iv) in Sect. 3.7, $\liminf A_j \subset \liminf A_{j,j}$. $\square$

Remark 3.45 By Theorem 3.28, a connected, locally compact metric space is σ-compact, and all σ-compact spaces are separable. Thus, every sequence $\{A_j\}$ of non-empty sets in a metric space which is either σ-compact, or connected and locally compact, has a convergent subsequence $\{A_{j_\ell}\}$.

If the metric space is compact, then $\lim A_{j_\ell} \neq \emptyset$, see (vii) in Sect. 3.7 and Theorem 3.44. Moreover, by Corollary 3.41, $\lim A_{j_\ell}$ is a continuum if all the A_j, $j \in \mathbb{N}$, are connected.

For extensions of Lemma 3.39 and Theorem 3.40 that apply to *unbounded* sets in a normed linear space when the upper limit set is non-empty (this is weaker than the lower limit set being non-empty), see [11–13].   □

3.9 Hausdorff Metric

Let $\mathcal{M}$ be the family of all non-empty, closed, bounded subsets of a metric space M and, for $A, B \in \mathcal{M}$, let

$$d_H(A, B) = \inf\{\delta > 0 : A \subset N_\delta(B) \text{ and } B \subset N_\delta(A)\}, \tag{3.15}$$

or equivalently

$$d_H(A, B) = \max\left\{ \sup_{a \in A} \text{dist}\,(a, B), \ \sup_{b \in B} \text{dist}\,(b, A) \right\}.$$

Clearly, $(\mathcal{M}, d_H)$ is a metric space, and d_H is called the Hausdorff metric [23, p. 42], [35, p. 281].

Lemma 3.46 $(\mathcal{M}, d_H)$ *is complete if* (M, d) *is complete.*

Proof If $\{X_k\}$ is a Cauchy sequence in $(\mathcal{M}, d_H)$ there is a subsequence, also denoted by $\{X_k\}$, such that $d_H(X_k, X_{k+1}) < 2^{-k}$ for all $k \in \mathbb{N}$. Then, for such a subsequence and every $\ell \in \mathbb{N}$,

$$X_\ell \cup X_{\ell+1} \subset N_{2^{-\ell}}(X_\ell), \quad X_\ell \cup X_{\ell+1} \cup X_{\ell+2} \subset N_{2^{-\ell}+2^{-\ell-1}}(X_\ell),$$

and hence by induction

$$\cup_{k \geqslant \ell} X_k \subset N_{2^{1-\ell}}(X_\ell) \quad \text{and} \quad \overline{\cup_{k \geqslant \ell} X_k} \subset N_{2^{2-\ell}}(X_\ell).$$

Moreover, for every $z \in X_\ell$, there exists a sequence $\{x_k\}_{k \geqslant \ell}$ such that $x_\ell = z$, $x_k \in X_k$ and $d(x_k, x_{k+1}) < 2^{-k}$ for all $k \geqslant \ell$; hence $\{x_k\}_{k \geqslant \ell}$ is Cauchy in M and converges to some x, because (M, d) is complete, with $d(x, z) < 2^{1-\ell}$.

Now if $X \subset M$ is the set of all $x \in M$ which are limits of these sequences (over all $\ell \in \mathbb{N}$ and $z \in X_\ell$), then X is non-empty, $X \subset N_{2^{2-\ell}}(X_\ell)$ and $X_\ell \subset N_{2^{1-\ell}}(X)$ for each $\ell \in \mathbb{N}$. It follows easily that $\overline{X} \in \mathcal{M}$ and $X_k \to \overline{X}$ in $(\mathcal{M}, d_H)$. This shows that the given Cauchy sequence has a subsequence converging in $(\mathcal{M}, d_H)$ to $\overline{X}$, and it follows that the Cauchy sequence converges in $(\mathcal{M}, d_H)$ to $\overline{X}$, as required.   □

Lemma 3.47 *If all bounded, closed subsets of M are compact, the set $C(M)$ of non-empty, closed, bounded, connected subsets of M is closed in* $(\mathcal{M}, d_H)$.

Proof Suppose $X_k \in C(M)$ and $X_k \to X$, in $(\mathcal{M}, d_H)$, where X is not connected in M. Since X is compact, there exist non-empty closed subsets Y_1 and Y_2, and

$\delta > 0$ with

$$X = Y_1 \cup Y_2, \qquad N_\delta(X) = N_\delta(Y_1) \cup N_\delta(Y_2) \text{ and } N_\delta(Y_1) \cap N_\delta(Y_2) = \emptyset.$$

Then for all k sufficiently large,

$$X_k \subset N_\delta(X) = N_\delta(Y_1) \cup N_\delta(Y_2) \text{ and } X_k \cap N_\delta(Y_1) \neq \emptyset \neq X_k \cap N_\delta(Y_2),$$

which contradicts X_k being connected. This completes the proof.   □

3.10   Components

Definition 3.48 A component of a set A is a connected set which is maximal by set inclusion (Remark 2.2) among all the connected subsets of A.   □

Remark 3.49 The empty set is a component of itself, but components of non-empty sets are non-empty. By Lemma 3.25, a component of a set A is closed in A, but may not be open in A; see Corollary 4.4 and Lemma 4.7.   □

The following lemma is motivated by [49, Ch. I, (13.1)].

Lemma 3.50 *Suppose $\emptyset \neq U \subset M$ where M is connected and the boundary ∂U is a subset of S where S has $n \in \mathbb{N}$ components. Then $S \cup U$ has no more than n components.*

Proof If $S = \emptyset$, U is open and closed because $\partial U = \emptyset$, and since U is non-empty, $U = M$ because M is connected. Hence S and $S \cup U$ each have one component, since M is connected and $S = \emptyset$ is connected by Remark 3.49.

So suppose $S \neq \emptyset$, and thus $n \geqslant 1$. The first step is to show that $m \leqslant n$, where n is the number of components of S and $\{F_j : 1 \leqslant j \leqslant m\}$ is any finite collection of non-empty disjoint sets that are closed in $S \cup U$ with $S \cup U = \cup_{j=1}^m F_j$. (Such a decomposition does exist, for example $m = 1$ and $F_1 = S \cup U$.) Then for all j, $1 \leqslant j \leqslant m$, $m \geqslant 1$, F_j is closed in $S \cup U$ and, since $S \cup U = \cup_{j=1}^m F_j$, it follows that each F_j is open and closed in $S \cup U$.

Therefore there exists a closed set A_j and an open set B_j in M such that

$$F_j = A_j \cap (S \cup U) \text{ and } F_j = B_j \cap (S \cup U).$$

Since $\partial U \subset S$, if $F_j \cap S = \emptyset$ for some $j \in \{1, \cdots, m\}$, $m \geqslant 1$, it follows that

$$F_j = A_j \cap (S \cup U) = A_j \cap (S \cup \overline{U}) = A_j \cap \overline{U},$$

and

$$F_j = B_j \cap (S \cup U) = B_j \cap (S \cup (U^\circ)) = B_j \cap (U^\circ).$$

This shows that $F_j \cap S = \emptyset$ implies that F_j is both open and closed in M, and hence $F_j = M$ for all such j, since M is connected. Thus, if such an F_j exists, it is unique, $m = 1$, and $m \leqslant n$ since $n \geqslant 1$.

Now suppose that $F_j \cap S \neq \emptyset$ for all $j \in \{1, \cdots, m\}$, $m \geqslant 1$. Since the components, E_i, $1 \leqslant i \leqslant n$, of S are connected in $S \cup U$, each E_i is a subset of exactly one of the sets F_j, $1 \leqslant j \leqslant m$, which are disjoint and closed in $S \cup U$. Moreover, $F_j \cap S \neq \emptyset$ for all j, $1 \leqslant j \leqslant m$, implies that for all j, $F_j \cap E_i \neq \emptyset$ for at least one i, $1 \leqslant i \leqslant n$. Hence $m \leqslant n$ in this case also.

Now let a family of closed sets $F_1, \cdots, F_m$ be as above with $m \geqslant 1$, and suppose that one of them, F_1 say, is not connected. Then there exist non-empty, disjoint subsets $\hat{F}_1$ and $\hat{F}_2$ closed in F_1 such that $F_1 = \hat{F}_1 \cup \hat{F}_2$. Hence $\{\hat{F}_1, \hat{F}_2, F_2, \cdots, F_m\}$ is also such a family of closed sets, but with $m + 1 \leqslant n$ elements.

Now let $\hat{m} \leqslant n$ be the largest integer for which there is a family of closed sets $F_1, \cdots, F_{\hat{m}}$ as described above. By maximality of $\hat{m}$, each F_j, $1 \leqslant j \leqslant \hat{m} \leqslant n$, is connected and thus a component of $S \cup U$, and the proof is complete. □

Remark 3.51 As a consequence, if a point x in a connected metric space M has an open neighbourhood U such that ∂U is a finite set, then both $\overline{U}$ and $M \backslash U$ have finitely many components, as observed in the proof of [49, Ch. I, (13.1)]. □

Lemma 3.52 *Suppose M, N are metric spaces, $f : M \to N$ is continuous and C is a component of N. Then $f^{-1}(C)$ is a union of components of M.*

Proof When $f(x) \in C$, let D be the component of M with $x \in D$. Then by Lemma 3.23, $f(D) \subset C$ since C is a component of N. The result follows. □

The next result, which in [40, Lemma 1.1] is attributed to Whyburn [50, p.12, (9.3)], is very important in the topological theory of global bifurcation which was summarised in Sect. 1.3 above.

Theorem 3.53 *When R and S are non-empty, disjoint, closed subsets of a compact set H in a metric space M and no component of H intersects both R and S, there exist closed sets H_r, H_s with*

$$H = H_r \cup H_s, \quad H_r \cap H_s = \emptyset, \quad R \subset H_r \text{ and } S \subset H_s, \tag{3.16}$$

and so H is not connected.

Proof Suppose there exists a sequence $0 < \epsilon_j \to 0$ such that for each $j \in \mathbb{N}$ there exist $r_j \in R$ and $s_j \in S$ joined in H by an ϵ_j-chain (see Definition 3.35), $A_j = \{x_i^{\epsilon_j} : 1 \leqslant i \leqslant n_{\epsilon_j}\} \subset H$ say. Since H is compact there is no loss in supposing that $r_j \to r \in R$ and $s_j \to s \in S$, and since $\{r, s\} \subset \liminf A_j \neq \emptyset$ it follows from Theorem 3.40 that $\limsup A_j$ is connected. But, since $\{r, s\} \subset \liminf A_j \subset \limsup A_j \subset H$, this contradicts the hypothesis that R and S are not joined by a continuum in H. This shows that for some $\epsilon^* > 0$ there is no ϵ^*-chain in H which joins a point of R to a point of S.

Now let H_r be the set of points of H which are joined to a point of R by an ϵ^*-chain and let $H_s = H \setminus H_r$. Then it follows from Lemma 3.37 that H_r and H_s are closed in H, and (3.16) follows. $\qquad \square$

Definition 3.54 If a sequence of distinct points of $X \subset M$ converges to $x \in M$, the point x is called a limit point of X, see [49, Ch. I, p. 2]. $\qquad \square$

Corollary 3.55 ([49, Ch. I, (10.1),(10.2)]) *For a connected set A and an open set G in M, suppose $\overline{G} \cap A$ is compact and $\emptyset \neq G \cap A \neq A$. Then*

(i) every component of $A \cap \overline{G}$ intersects ∂G;
(ii) every component of $A \cap G$ has a limit point in ∂G.

Proof (i) Let $H = A \cap \overline{G}$, $S = A \cap \partial G$ and let R be a component of H for which $R \cap \partial G = \emptyset$. Then H is compact by hypothesis, $S = H \cap \partial G$ is closed, S is nonempty because otherwise $A = (A \cap G) \cup (A \setminus \overline{G})$ would separate A, and $R \neq \emptyset$ is closed because it is a component of H.

Since $R \cap S = \emptyset$, by Theorem 3.53 there are closed sets H_r, H_s such that (3.16) holds. Moreover, $A \cap \partial G = S \subset H_s$ and $H_r = H \setminus H_s$ implies that $H_r \subset A \cap G$. Therefore the two closed subsets of M,

$$H_r \quad \text{and} \quad H_s \cup (M \setminus G),$$

are disjoint and cover A, which is false since A is connected. This proves (i), that every component of H intersects ∂G.

(ii) Let T be a component of $A \cap G$ for which no limit point of T belongs to ∂G. Then if T is not closed in M there exists $x_k \to x \notin T$ where $\{x_k\} \subset T$, and since this implies that x is a limit point of T, $x \notin \partial G$. Since $x_k \in \overline{G} \cap A$ which is compact, it follows that $x \in A \cap G$ and hence $x \in T$ because, by Lemma 3.25, T is closed in $A \cap G$. Since this is false, T is closed in M, and hence T is compact in M since it is a subset of the compact set $\overline{G} \cap A$.

Therefore, since T and $M \setminus G$ are disjoint and closed, there is an open set O in M such that $T \subset O \subset \overline{O} \subset G$. Then by (i), $C \cap \partial O \neq \emptyset$ where C is the component of $A \cap \overline{O}$ which contains T. This implies that $T \subset C$ but $C \neq T$ because $T \subset O$. However this is false because C is connected, $C \subset A \cap \overline{O} \subset A \cap G$ and T is a component of $A \cap G$. This contradiction proves that every component T of $A \cap G$ has a limit point in ∂G, as required. $\qquad \square$

3.11   Composants

Definition 3.56 When K is a non-degenerate continuum and $p \in K$, the union of all proper sub-continua of K which contain p is called the composant of p and denoted by K_p, see [23, pp. 139–140] or [37, Ch. XI]. $\qquad \square$

Theorem 3.57 *When K is a non-degenerate continuum and $p \in K$,*

(i) K_p is connected and dense in K, and $p \in K_p$;
(ii) K_p is the union of countably many proper sub-continua of K which contain p;
(iii) if $K = K' \cup K''$ where K' and K'' are proper sub-continua of K, then $K' \cap K'' \neq \emptyset$ and $K = K_p$ for all $p \in K' \cap K''$.

Proof From Definition 3.21, K can be regarded as a non-degenerate, compact, connected metric space.

(i) As $\{p\}$ is a proper sub-continuum of K, $p \in K_p$. To show that K_p is connected, suppose it is not. Then there exist sets G_1 and G_2 which are open in K with

$$K_p \cap G_1 \cap G_2 = \emptyset, \quad K_p \subset G_1 \cup G_2, \quad G_1 \cap K_p \neq \emptyset \neq K_p \cap G_2.$$

Suppose $p \in G_1$. Then if $p \in K'$ where K' is a proper sub-continuum of K, it is immediate that $K' \subset G_1$, whence $K_p \subset G_1$ and $G_2 \cap K_p = \emptyset$. This contradiction shows that K_p is connected for all $p \in K$.

To show that K_p is dense in K it suffices to show that $K_p \cap U \neq \emptyset$ for every non-empty open set U in K. So let U be open in K and let B be an open ball with $\emptyset \neq \overline{B} \subset U$. Then if $p \in \overline{B}$ there is nothing to prove.

So suppose $p \notin \overline{B}$ and let W be the component of $K \setminus \overline{B}$ that contains p. Then by Lemma 3.25 $\overline{W}$ is a proper connected subset of K which contains p, and hence $\overline{W} \subset K_p$.

Now with $M = A = K$ and $G = K \setminus \overline{B}$ in Corollary 3.55 (ii), it follows that $\overline{W} \cap \partial(K \setminus \overline{B}) \neq \emptyset$ which, since $\partial(K \setminus \overline{B}) = \partial \overline{B} \subset \overline{B} \subset U$, shows that

$$\emptyset \neq \overline{W} \cap \partial(K \setminus \overline{B}) \subset K_p \cap U.$$

Since $U \cap K_p \neq \emptyset$ for all non-empty open sets U in K, this proves (i).

(ii) Because K is compact, for fixed $p \in K$ there exists a countable base $\{V_i \neq \emptyset : i \in \mathbb{N}\}$ of $K \setminus \{p\}$ with $p \notin \overline{V}_i$ for all i. Let C_i, $i \in \mathbb{N}$, be the component of $K \setminus \overline{V}_i$ which contains p. Then $\overline{C}_i \subset K_p$, because $p \in \overline{C}_i$, a proper connected sub-continuum of K. This shows that $\bigcup_{i \in \mathbb{N}} \overline{C}_i \subset K_p$.

Now for any $q \in K_p$ there is a proper sub-continuum K' of K with $\{p, q\} \subset K'$. Fix $r \in K \setminus K'$. Then since $\{V_i\}$ is a base of $K \setminus \{p\}$, there exists $j \in \mathbb{N}$ with $r \in V_j$ and $\overline{V}_j \subset K \setminus K'$; equivalently $K' \subset K \setminus \overline{V}_j$ which implies that $K' \subset C_j$ since $p \in K'$. Since $q \in K'$ it follows that $q \in C_j$. Hence

$$K_p \subset \bigcup_{i \in \mathbb{N}} C_i \subset \bigcup_{i \in \mathbb{N}} \overline{C}_i \subset K_p,$$

where $p \in \overline{C}_i$, a proper connected sub-continuum of K; this proves (ii).

(iii) If $K = K' \cup K''$, there exists $p \in K' \cap K''$ because K is connected. Then since K' and K'' are proper sub-continua of K, by Definition 3.56,

$$K = K' \cup K'' \subset K_p \subset K \text{ and the result follows.} \qquad \square$$

Remark 3.58 For composants of indecomposable continua, Definition 6.1, see Theorem 6.5. $\qquad \square$

Chapter 4
Types of Connectedness

4.1 Path-Connected Sets

Definition 4.1 For $A \subset M$, the set $\{f(t) : t \in [0, 1]\}$, where $f : [0, 1] \to A$ is continuous with $f(0) = x$ and $f(1) = y$, is a path in A joining x to y. A set A is path-connected if all pairs $x, y \in A$ are joined by a path in A, and a path-component of A is maximal by set inclusion among all path-connected subsets of A. □

Lemma 4.2 *An open connected set A in an normed linear space is path-connected.*

Proof If M is a normed linear space, by Corollary 3.34(b) let $B_{\epsilon_i}(x_i) \subset A$ and $\epsilon_i > 0$ be such that $x_1 = x$, $x_n = y$ and

$$B_{\epsilon_i}(x_i) \cap B_{\epsilon_j}(x_j) \neq \emptyset \text{ if and only if } |i - j| \leq 1, \quad 1 \leq i, j \leq n.$$

Then let $L_i = \{(1-t)x_i + tx_{i+1} : t \in [0, 1]\}$, $1 \leq i \leq n-1$, be straight line segments in M joining consecutive points x_i. Then clearly $L_i \subset B_{\epsilon_i}(x_i) \cup B_{\epsilon_{i+1}}(x_{i+1}) \subset A$, $L_i \cap L_{i+1} = \{x_{i+1}\}$, $1 \leq i \leq n - 1$, and $L := \cup_{i=1}^{n-1} L_i$ is a piecewise-affine path in A joining x to y. □

Theorem 4.3 *A path-connected set A is connected.*

Proof If A is not connected there exist non-empty sets H_1 and H_2, both of which are open and closed in A, with $A = H_1 \cup H_2$ and $H_1 \cap H_2 = \emptyset$. Then if A is path-connected, for $x \in H_1$ and $y \in H_2$ there exists a continuous function $f : [0, 1] \to A$ with $f(0) = x$ and $f(1) = y$, and the sets $I_j = f^{-1}(H_j)$, $j = 1, 2$, are non-empty, open and closed in $[0, 1]$, and $[0, 1] = I_1 \cup I_2$. Hence, since $[0, 1]$ is connected by Remark 3.22, there exists $t \in I_1 \cap I_2$, in which case $f(t) \in H_1 \cap H_2$, a contradiction. □

Corollary 4.4 *A component of an open set in a normed linear space is open.*

© The Author(s), under exclusive license to Springer Nature Switzerland AG 2025
B. Buffoni, J. Toland, *Connected Sets in Global Bifurcation Theory*, SpringerBriefs in Mathematics, https://doi.org/10.1007/978-3-031-87051-4_4

Proof Suppose $x \in U$ where U is a component of G and G is open in a normed linear space. Then $B_\eta(x) \subset G$ for some $\eta > 0$ and, since it is path-connected by Lemma 4.2, $B_\eta(x)$ is connected. Since U is a component of G, it follows that $B_\eta(x) \subset U$, and hence U is open.   □

The converse of Theorem 4.3 is false, as the following example illustrates.

Example 4.5 (The Topologist's Sine Curve)   Figure 4.1 below represents $\overline{T} = T \cup (\{0\} \times [-1, 1])$ which is the closure in $\mathbb{R}^2$ of the graph $T = \{(t, \sin(1/t)) : t > 0\}$. Since T is path-connected, it is connected by Theorem 4.3, and so $\overline{T}$ is connected by Lemma 3.25. But there is no path in $\overline{T}$ joining $(1/\pi, 0)$ to $(0, 0)$ because no function $f : [0, 1] \to \overline{T} \subset \mathbb{R}^2$ with $f(0) = (0, 0)$ and $f(1) = (1/\pi, 0)$ can be continuous. To see this suppose such a function $f = (f_1, f_2)$ exists. Then

$$f_1(s) > 0 \text{ implies } f_2(s) = \sin(1/f_1(s)),$$

$$f_1(0) = 0, \ f_1(1) = 1/\pi,$$

and by the intermediate value theorem there exists $s_1 \in (0, 1)$ with $f_1(s_1) = \frac{2}{3\pi}$, and $s_2 \in (0, s_1)$ with $f_1(s_2) = \frac{2}{5\pi}$, and by induction

$$s_k \in (0, s_{k-1}) \text{ with } f_1(s_k) = \frac{2}{(2k+1)\pi}$$

for all $k \in \mathbb{N}$, where $s_0 = 1$. Therefore

$$f_2(s_k) = \sin \frac{(2k+1)\pi}{2} = (-1)^k, \ k \in \mathbb{N}.$$

Moreover, $s_k \searrow s_* \in [0, 1)$ and since f is continuous it follows that $(-1)^k = f_2(s_k) \to f_2(s_*)$, which is false. Hence $\overline{T}$ is not path-connected.

Fig. 4.1 The topologist's sine curve in $\mathbb{R}^2$

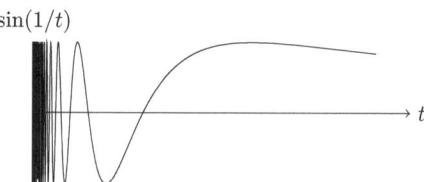

4.2   Locally Connected Sets

Definition 4.6 A metric space M is locally connected at x if every neighbourhood of x contains a connected open neighbourhood of x, and M is locally connected if it is locally connected at every point of M [23, § 3-1].   □

A non-empty set $A \subset M$ is locally connected at $a \in A$ if (A, d_A) is locally connected at a, A is locally connected if it is locally connected at every point of A, and the set $A = \emptyset$ will be considered as locally connected.

Remark A locally connected set need not be connected. For example, $A = (-1, 0) \cup (0, 1) \subset \mathbb{R}$ is locally connected but not connected. □

Definition 4.6 is standard for local connectedness at a point but it is not the one used by Whyburn [49, 50]. For more about the distinction, see Definition 4.21, Remark 4.22, and Lemma 4.23. For the following results, see [23, Thm. 3-2 and Lemma 3-21]. □

Lemma 4.7 *A set A is locally connected if and only if when U is open in A all the components of U are open in A.*

Proof It suffices to consider the case $A = M$. So suppose M is locally connected, let $U \subset M$ be open and let $x \in C$ where C is a component of U. Since U is a neighbourhood of x, there is an open connected set V with $x \in V \subset U$, and since C is a component of U, it follows that $V \subset C$, and hence C is open.

Conversely suppose that all the components of any open set U in M are open. Then let $x \in M$, let V be a neighbourhood of x and let U be an open set with $x \in U \subset V$. It follows that the component C of U with $x \in C$ is an open connected set with $x \in C \subset U \subset V$. Thus M is locally connected since V was an arbitrary neighbourhood of an arbitrary point $x \in M$. □

Lemma 4.8 *If a metric space M is locally connected at x, then every open subset A containing x is locally connected at x.*

Proof Suppose $x \in A$ where A is open in M, and let $N \subset A$ be a neighbourhood of x in (A, d_A). Then there exists U which is open in A with $x \in U \subset N \subset A$. Then since A is open, U is open in M and therefore U is a neighbourhood of x in M. Now since M is locally connected at x, there exists a connected open set V in M with $x \in V \subset U \subset N \subset A$. Since the connected set $V \subset N$ is open in (A, d_A), the result follows. □

Corollary 4.9 *An open set in a locally connected space is locally connected.*

Proof This is immediate from Lemma 4.8. □

Lemma 4.10 *Suppose M is locally connected, N is a metric space, $f : M \to N$ is continuous and $f(H)$ is closed in $f(M)$ (with the metric inherited from N) when H is closed in M. Then $f(M)$ is locally connected.*

Remark If M is compact and $H \subset M$ is closed, then $f(H)$ is closed in N, and thus in $f(M)$. □

Proof Let U be open in $f(M)$ and C be a component of U. For $x \in f^{-1}(C)$ let K_x be the component of $f^{-1}(U)$ which contains x. Then K_x is open by Lemma 4.7, $x \in K_x$ and $f(K_x)$ is connected. Since $f(x) \in C$ it follows that $f(K_x) \subset C$ and hence that $x \in K_x \subset f^{-1}(C)$.

This shows that $f^{-1}(C)$ is open, and hence that $f^{-1}(N \setminus C) = M \setminus f^{-1}(C)$ is closed in M. Now, since f maps closed sets in M to closed sets in $f(M)$,

$$f\big(f^{-1}(N \setminus C)\big) = f\big(f^{-1}(f(M) \setminus C)\big) = f(M) \setminus C$$

is closed in $f(M)$. This shows that every component C of every open subset U of $f(M)$ is open in $f(M)$, and by Lemma 4.7 the result follows.   □

4.3   Peano Continua

Definition 4.11 A Peano continuum (generalised Peano continuum) is a locally connected continuum (locally connected generalised continuum). See Definition 3.21.   □

Example Since the compact interval $[0, 1]$ is locally connected it is a Peano continuum, and hence any path in a metric space is a Peano continuum, by Lemma 4.10.   □

Theorem 4.12 *A generalised Peano continuum is path-connected.*

Proof The aim is to show that a generalised Peano continuum M has a path joining x to y for all $x \neq y \in M$.   □

Definition 4.13 Let $V_1, \ldots, V_n$, $n \in \mathbb{N}$, be non-empty, open, connected subsets of M and let $I_1, \ldots, I_n$ be non-empty, disjoint open intervals in $[0, 1]$ such that $\cup_{j=1}^n \overline{I_j} = [0, 1]$ and $\sup I_j = \inf I_{j+1}$, $j \in \{1, \ldots, n-1\}$ if $n \geq 2$. Then let

$$K_j = \overline{I_j} \times \overline{V_j} \text{ and } [\mathscr{K}] = \cup_{j=1}^n K_j \text{ where } \mathscr{K} = \{K_1, \ldots, K_n\}.$$

If $x \in V_1$, $y \in V_n$ and $V_j \cap V_{j+1} \neq \emptyset$ for all $j \in \{1, \ldots, n-1\}$ (no condition if $n = 1$), $\mathscr{K}$ will be called a thick path with n pieces joining x and y, and a compact thick path if $[\mathscr{K}]$ is compact. Then by Remark 3.19 and Lemma 3.20 $[\mathscr{K}]$ is closed and connected in $\mathbb{R} \times M$, and its thickness can be defined by

$$\text{T}(\mathscr{K}) = \sup \big\{ d(u, v) : (s, u), (s, v) \in [\mathscr{K}], \ s \in [0, 1] \big\} \in [0, \infty].$$ □

Lemma 4.14 *For a thick path $\mathscr{K}$ with n pieces joining x to y in a generalised Peano continuum there is a thick path $\widetilde{\mathscr{K}}$ with $\widetilde{n} \geq n$ pieces joining x to y, such that*

$$[\widetilde{\mathscr{K}}] \text{ is compact, } [\widetilde{\mathscr{K}}] \subset [\mathscr{K}] \text{ and } \text{T}(\widetilde{\mathscr{K}}) \leq \tfrac{1}{2} \min \big\{ \text{T}(\mathscr{K}), 1 \big\}.$$

Proof Let $\delta = \min\{\text{T}(\mathscr{K}), 1\}$. If $\delta = 0$ each V_j is a singleton and since $\cup_1^n V_j$ is connected they are all the same singleton, which is false since $x \in V_1$, $y \in V_n$ and $x \neq y$. So with $\delta > 0$, let

$x_0 = x \in V_1$, $x_n = y \in V_n$ and $x_j \in V_j \cap V_{j+1}$, $1 \leqslant j \leqslant n-1$ (if $n \geqslant 2$).

Since M is locally connected, V_j is locally connected by Corollary 4.9, and since M is locally compact, V_j is locally compact, for each j. Therefore, by Theorem 3.33, there is a simple chain $\mathcal{V}^j$ which joins x_{j-1} to x_j, with open connected links having compact closures in V_j and diameters less than $\delta/4$.

Write $\mathcal{V}^j = \{V_1^j, \cdots, V_{N_j}^j\}$, $N_j \in \mathbb{N}$, and for each $j \in \{1, \ldots, n\}$ let

$$\overline{I_j} = \overline{I_1^j} \cup \ldots \cup \overline{I_{N_j}^j},$$

where $I_1^j, \ldots, I_{N_j}^j$ are non-empty open intervals in I_j such that $\sup I_i^j = \inf I_{i+1}^j$ for all $i \in \{1, \ldots, N_j - 1\}$ (if $N_j \geqslant 2$), and let

$$\widetilde{K_i^j} = \overline{I_i^j} \times \overline{V_i^j} \text{ for } j \in \{1, \ldots, n\} \text{ and } i \in \{1, \ldots, N_j\}.$$

Then

$$\mathcal{K} = \left\{ \widetilde{K_i^j} : 1 \leqslant j \leqslant n, \ 1 \leqslant i \leqslant N_j \right\} = \left\{ \widetilde{K_1^1}, \ldots, \widetilde{K_{N_1}^1}, \ldots, \widetilde{K_1^n}, \ldots, \widetilde{K_{N_n}^n} \right\}$$

is a thick path joining x to y which satisfies the desired properties with $\widetilde{n} = N_1 + \ldots + N_n$ pieces.   □

To complete the proof of Theorem 4.12, consider the thick path $\mathcal{K}_0 = \{[0, 1] \times M\}$ joining x and y. Then by Lemma 4.14 there is a compact thick path $\mathcal{K}_1$ joining x and y, such that $T(\mathcal{K}_1) \leqslant \frac{1}{2}$ and by induction there is a sequence $\{\mathcal{K}_\ell\}_{\ell \in \mathbb{N}}$ of compact thick paths joining x and y with

$$T(\mathcal{K}_\ell) \leqslant 2^{-\ell} \text{ and } [\mathcal{K}_1] \supset [\mathcal{K}_2] \supset [\mathcal{K}_3] \supset \ldots.$$

Therefore, by Theorem 3.43, $P := \bigcap_{\ell \geqslant 1}[\mathcal{K}_\ell]$ is a continuum in $[0, 1] \times M$ that contains $(0, x)$ and $(1, y)$, and such that

$$\sup \left\{ d(u, v) : \text{ there exists } s \in [0, 1] \text{ such that } (s, u), (s, v) \in P \right\} = 0.$$

Now if $s \in (0, 1)$ is such that $(\{s\} \times M) \cap P = \emptyset$, then

$$P = \left(([0, s) \times M) \cap P \right) \bigcup \left(((s, 1] \times M) \cap P \right)$$

is a separation of P, since $(0, x) \in ([0, s) \times M) \cap P$ and $(1, y) \in ((s, 1] \times M) \cap P$. Since P is connected it follows that $(\{s\} \times M) \cap P$ is non-empty and hence a singleton, $\{(s, f(s))\}$ say, for each $s \in [0, 1]$. In particular, $f(0) = x$ and $f(1) = y$.

Finally, to see that $f : [0, 1] \to M$ is continuous, suppose $s_j \to s$ as $j \to \infty$, $s_j \in [0, 1]$, but $d(f(s_j), f(s)) \not\to 0$, and without loss assume that $d(f(s_j), f(s)) \geqslant \delta > 0$ for all $j \in \mathbb{N}$. Now since $(s_j, f(s_j)) \in P$ and P is compact, a subsequence $\{(s_{j_k}, f(s_{j_k}))\}$ converges in P and, since $s_{j_k} \to s$, it follows that its limit is $(s, f(s)) \in P$. But this is false since $d(f(s_j), f(s)) \geqslant \delta > 0$ for all $j \in \mathbb{N}$. Hence f is continuous and $P = \{(s, f(s)) : s \in [0, 1]\}$ is a path joining x to y in M.

Theorem 4.12 is key in proving the following important result which is cited without proof because Theorem 4.12 suffices for the subsequent chapters.

Hahn-Mazurkiewicz Theorem
(a) $A \subset M$ is a Peano continuum if and only if there exists a continuous function $f : [0, 1] \to M$ such that $A = f([0, 1])$.
(b) A is a generalised Peano continuum only if there exists a continuous function $h : [0, \infty) \to M$ such that $A = h([0, \infty))$. The converse is false.

Proof For (a), see Hahn [20] and Mazurkiewicz [32], and many other sources such as [17, 23, 32, 49, 51, 52]. Mazurkiewicz, after his proof of (a), showed that (b) holds but the converse is false. For detailed theory of (b), see [1].   □

4.4 Arc-Connectedness

Definition 4.15 When f in Definition 4.1 is injective, the path joining x to y is called an arc and A is said to be arc-connected if every pair of distinct points in A are joined by an arc in A.   □

Since there is no such thing as an arc joining x to itself, suppose for the rest of this section that $x \neq y$.

If f defines a path $P = \{f(t) : t \in [0, 1]\}$ that is not an arc there exist $s \neq t \in [0, 1]$ such that $f(s) = f(t)$, and a path may have uncountably many such points of self-intersection. Nevertheless, for any path P joining x to y, there is an arc $\tilde{P} : \{\tilde{f}(t) : t \in [0, 1]\} \subset P$ joining x to y. Indeed it is well known that this is true in a Hausdorff topological space [52, p. 197]. There follows a direct proof that a generalised Peano continuum is arc-connected.

Lemma 4.16 *When $\{a, b\} \subset V$, a connected open subset of a locally connected metric space M, there exists a connected open subset V^{ab} of M such that $\{a, b\} \subset \overline{V^{ab}} \subset V$.*

Proof For $a \in V$ let V^a be the set of $b \in V$ such that such a V^{ab} exists. Then since $a \in V^a$ because M is locally connected, and since V^a is open and closed, $V^a = V$ for all $a \in V$, because V is connected.   □

Corollary 4.17 *In a locally connected metric space M let $\{V_1, \ldots, V_n\}$, $n \geqslant 2$, be a simple chain with connected open links joining x to y with*

$$x_0 = x, \quad x_n = y \text{ and } x_j \in V_j \cap V_{j+1} \text{ for all } j \in \{1, \dots, n-1\}. \tag{4.1}$$

Then there exists a simple chain $\{W_1, \dots, W_n\}$ with open connected links joining x to y such that $\overline{W_j} \subset V_j$, $j \in \{1, \dots, n\}$, and $x_j \in W_j \cap W_{j+1}$, $j \in \{1, \dots, n-1\}$.

Proof Let $W_j = V_j^{x_{j-1}x_j}$ in the notation of Lemma 4.16. $\qquad\qquad\square$

Definition 4.18 In the terminology of Definition 4.13, a thick path $\mathcal{K} = \{K_1, \dots, K_n\}$ with $K_j = \overline{I_j} \times \overline{V_j}$ $(1 \leqslant j \leqslant n)$ is called a *thick arc* if when $n \geqslant 2$ it also satisfies

$$x \notin \overline{V_2}, \quad y \notin \overline{V_{n-1}} \text{ and } |j_1 - j_2| \geqslant 2 \text{ implies that dist } (V_{j_1}, V_{j_2}) > 0. \tag{4.2}$$

Let $\mathrm{L}(\mathcal{K}) > 0$ be the maximum of the lengths of $I_1, \dots, I_n$. $\qquad\qquad\square$

The following is an analogue of Lemma 4.14 for thick arcs.

Lemma 4.19 *For a thick arc $\mathcal{K}$ with n pieces joining x to y in a generalised Peano continuum there is a thick arc $\widetilde{\mathcal{K}}$ with $\tilde{n} \geqslant 2n$ pieces joining x to y, such that $[\widetilde{\mathcal{K}}]$ is compact,*

$$[\widetilde{\mathcal{K}}] \subset [\mathcal{K}], \ \mathrm{T}(\widetilde{\mathcal{K}}) \leqslant \tfrac{1}{2} \min \{\mathrm{T}(\mathcal{K}), 1\} \text{ and } \mathrm{L}(\widetilde{\mathcal{K}}) \leqslant \tfrac{1}{2}\mathrm{L}(\mathcal{K}).$$

Proof For the thick arc

$$\mathcal{K} = \{K_1, \dots, K_n\} = \{\overline{I_1} \times \overline{V_1}, \dots, \overline{I_n} \times \overline{V_n}\}$$

let $\delta = \min\{T(\mathcal{K}), 1\}$ and note that, when $n \geqslant 2$, (4.1) holds by Definition 4.13, and (4.2) holds by Definition 4.18. Since for each $j \in \{1, \dots, n\}$, V_j is an open generalised Peano continuum, by Theorem 3.33 applied to an appropriate open cover of V_j, there is a simple chain

$$\mathcal{V}_j = \{V_1^j, \cdots, V_{N_j}^j\} \text{ joining } x_{j-1} \text{ to } x_j \text{ in } V_j \text{ with diam } (V_i^j) < \tfrac{1}{4}\delta,$$

where the links V_i^j are open, connected and have compact closure in V_j with the following properties:

$$\begin{aligned}
\text{diam } (V_i^j) &< \tfrac{1}{2}\text{dist}(V_{j-1}, V_{j+1}) && \text{if } 1 < j < n; \\
\text{diam } (V_i^j) &< \tfrac{1}{2}\text{dist}(x, V_2) && \text{if } j = 1 < n; \\
\text{diam } (V_i^j) &< \tfrac{1}{2}\text{dist}(y, V_{n-1}) && \text{if } j = n > 1; \\
\text{diam } (V_i^j) &< \tfrac{1}{2}d(x, y) && \text{if } j = n = 1.
\end{aligned}$$

Now when $n \geqslant 2$, irrespective of the intermediate points $x_1, \ldots, x_{n-1}$ some of the V_i^j can be deleted so that the remaining links V_i^j form a simple chain in M joining x to y.

To be precise, with $k_1 = 1$ and $i_n = N_n$, let i_1 be the smallest index $i \in \{k_1, \ldots, N_1\}$ such that $V_i^1 \cap V_{k_2}^2 \neq \emptyset$ for some $k_2 \in \{1, \ldots, N_2\}$, and for that i_1 choose k_2 to be the largest. Then let i_2 be the smallest index $i \in \{k_2, \ldots, N_2\}$ such that $V_i^2 \cap V_{k_3}^3 \neq \emptyset$ for some $k_3 \in \{1, \ldots, N_3\}$, and for that i_2 let k_3 be the largest. Because $\mathscr{K}$ is a thick arc, repeating this step eventually leads to sets of indices $i_1, \ldots, i_n$ and $k_1, \ldots, k_n$ such that

$$\{V_{k_1}^1, \ldots, V_{i_1}^1, V_{k_2}^2, \ldots, V_{i_2}^2, \ldots, V_{k_n}^n, \ldots, V_{i_n}^n\}$$

is a simple chain joining x and y.

If $n = 1$, $\mathscr{V}_1 = \{V_1^1, \ldots, V_{N_1}^1\}$ is a simple chain joining x to y where $N_1 \geqslant 2$, because otherwise $0 < d(x, y) \leqslant \text{diam}(V_1^1) \leqslant d(x, y)/2$, which is false. Since no link V_j^1 can be deleted, let $k_1 = 1$ and $i_1 = N_1$.

Also for all $n \geqslant 2$, $\{V_{k_1}^1, \ldots, V_{i_1}^1\}$ has at least two links because otherwise $0 < \text{dist}(x, V_2) \leqslant \text{diam}(V_1^1) = \text{diam}(V_{k_1}^1) \leqslant \text{dist}(x, V_2)/2$, which is false. For the same reason the simple chain $\{V_{k_n}^n, \ldots, V_{i_n}^n\}$ has at least two links.

Moreover, if $1 < j < n$, $\{V_{k_j}^j, \ldots, V_{i_j}^j\}$ has at least two links because otherwise $0 < \text{dist}(V_{j-1}, V_{j+1}) \leqslant \text{diam}(V_{k_j}^j) \leqslant \text{dist}(V_{j-1}, V_{j+1})/2$, which is false.

Now, for each $j \in \{1, \ldots, n\}$ and $i \in \{k_j, \ldots, i_j\}$, let I_i^j be an open interval of length depending only on j (but not on i), so that $\overline{I_j} = \overline{I_{k_j}^j} \cup \ldots \cup \overline{I_{i_j}^j}$ and $\sup I_i^j = \inf I_{i+1}^j$ for all $i \in \{k_j, \ldots, i_j - 1\}$. Let

$$\mathscr{K} = \left\{\overline{V_{k_1}^1} \times \overline{I_{k_1}^1}, \ldots, \overline{V_{i_1}^1} \times \overline{I_{i_1}^1}, \ldots, \overline{V_{k_n}^n} \times \overline{I_{k_n}^n}, \ldots, \overline{V_{i_n}^n} \times \overline{I_{i_n}^n}\right\}.$$

Decreasing each link in $\left\{V_{k_1}^1, \ldots, V_{i_1}^1, \ldots, V_{k_n}^n, \ldots, V_{i_n}^n\right\}$ by applying Corollary 4.17, but keeping the same notations for the decreased links, $\widetilde{\mathscr{K}}$ becomes a thick arc joining x to y which satisfies the desired properties.   □

From the Remark following Definition 4.15 the next result is immediate by Theorem 4.12, but the proof below is self-contained. For alternative self-contained proofs see [23, [Thms.3-15,-16,-17], [49, Ch. II, (5.1)], [52, p. 220].

Theorem 4.20 *A generalised Peano continuum is arc-connected.*

Proof First note that for all $\ell \geqslant 0$ the thick paths $\mathscr{K}_\ell$ in the proof of Theorem 4.12 may, by Lemmas 4.16 and 4.19, be regarded as thick arcs with $\text{L}(\mathscr{K}_\ell) \leqslant 2^{-\ell}$. So assuming that this is true for all ℓ the aim is to show that the continuous map f constructed in the proof of Theorem 4.12 is injective; in other words it defines an arc joining x to y.

So suppose this is false and $f(s_1) = f(s_2)$, z say, with $0 \leqslant s_1 \leqslant s_2 \leqslant 1$. Then (s_1, z) and (s_2, z) belong to all thick arcs $\mathcal{K}_\ell$, $\ell \geqslant 0$. For each $\ell \geqslant 0$, write

$$\mathcal{K}^\ell = \{\widehat{K}_1^\ell, \ldots, \widehat{K}_{m_\ell}^\ell\} \text{ with } \widehat{K}_j^\ell = \overline{\widehat{I}_j^\ell} \times \overline{\widehat{V}_j^\ell} \text{ and } m_\ell \in \mathbb{N}.$$

Suppose that (s_1, z) and (s_2, z) belong respectively to the pieces $\overline{\widehat{I}_{i_1}^\ell} \times \overline{\widehat{V}_{i_1}^\ell}$ and $\overline{\widehat{I}_{i_2}^\ell} \times \overline{\widehat{V}_{i_2}^\ell}$. Then since $\mathrm{dist}(\widehat{V}_{i_1}^\ell, \widehat{V}_{i_2}^\ell) = 0$ and $\mathrm{length}(\widehat{I}_j^\ell) \leqslant 2^{-\ell}$ for all $j \in \{1, \ldots, m_\ell\}$, and since $\mathcal{K}_\ell$ is a thick arc, it follows from (4.2) that $|i_1 - i_2| \leqslant 1$ and thus $s_2 - s_1 \leqslant 2^{(1-\ell)}$. Since $\ell \geqslant 0$ can be arbitrarily large, $s_1 = s_2$, which shows that f is injective. $\qquad\square$

4.5 Weak Local Connectedness

Definition 4.21 A metric space M is weakly locally connected at x if every neighbourhood of x contains a connected neighbourhood of x. If M is weakly locally connected at all its points, M is weakly locally connected. By Lemma 3.25 there is no loss in saying M is weakly locally connected at x if every neighbourhood of x contains a closed connected neighbourhood of x. $\qquad\square$

A non-empty subset A of M is weakly locally connected if (A, d_A) is weakly locally connected, and the set $A = \emptyset$ will be considered as weakly locally connected.

Remark 4.22 Although Definition 4.6 obviously implies Definition 4.21, the latter does not require the connected neighbourhood of x to be open, and they are not equivalent. For example the set $\mathcal{B}$ in Fig. 5.1 is weakly locally connected, but not locally connected, at $(0, 0) \in \mathcal{B}$, see [44, p. 139]. This is a subtle issue because for metric spaces the two definitions are equivalent, as is shown in the following Lemma. $\qquad\square$

Lemma 4.23 *A metric space is weakly locally connected if and only if it is locally connected.*

Proof Since Definition 4.6 implies Definition 4.21, it suffices to prove the converse using the criterion in Lemma 4.7.

Let M be a weakly locally connected metric space and let $U \neq \emptyset$ be an open subset of M. Then let C be a component of U and let $x \in C$. Since U is a neighbourhood of x, by the weak local connectedness of M, x has a connected neighbourhood $V_x \subset U$, and it follows that $V_x \subset C$ since C is a component of U. Now, since V_x is a neighbourhood of x there is an open set W with $x \in W \subset V_x \subset C$, which shows that C is open. Since an arbitrary component of an arbitrary open set is open, M is locally connected by Lemma 4.7. $\qquad\square$

Remark 4.24 If M is not weakly locally connected at x, it is not locally connected at x. But if M is not locally connected at y, although M is not weakly locally connected by Lemma 4.23, it may, or may not, be weakly locally connected at y. For example, as noted in Remark 4.22, $M = \mathscr{B}$ in Fig. 5.1 with $y = (0, 0)$.

Definition 4.21, of weak local connection at a point, is what Whyburn [49, p. 18] refers to as local connectedness at a point. So the difference between Definitions 4.6 and 4.21 at a point should be kept in mind when comparing [49] with other sources.

$\square$

The next theorem is related to the notion of semi-local-connectedness [49, Ch. I, (13.1)]; see also [37, Ex. 8.44, p. 136], [51, p. 233 & Rmk. 4.12 on p. 333], and [23, Thm. 3-19].

Theorem 4.25 *Suppose M is a weakly locally connected, generalised continuum. Then for every $p \in M$ and $\epsilon > 0$ there is a closed neighbourhood V of p with* $diam\,(V) < \epsilon$ *and $M \setminus V$ has finitely many components.*

Proof By Lemma 4.23, M is a generalised Peano continuum and hence path-connected by Theorem 4.12. For convenience, as in (3.1b), let $cA = M \setminus A$.

First, for $p \in M$ and $\epsilon > 0$, suppose there exists $\delta \in (0, \epsilon)$ such that $cB_\epsilon(p)$ is a subset of the union, U say, of finitely many components of $c(\overline{B_\delta(p)})$. Then $V := cU$ is a neighbourhood of p with $\overline{B_\delta(p)} \subset V \subset B_\epsilon(p)$. Since $U = M \setminus V$ has finitely many components, the result holds in this case. Indeed, as M is locally connected, by Lemma 4.23, each component of the open subset $c(\overline{B_\delta(p)})$ is open by Lemma 4.7, and hence U is open. Thus V is closed.

Therefore, for all $p \in M$, it is sufficient to show that there exist a sequence $\epsilon_j \searrow 0$ and $\delta_j \in (0, \epsilon_j)$ for all j, such that $cB_{\epsilon_j}(p)$ is a subset of the union U_j of finitely many components of $c(\overline{B_{\delta_j}(p)})$, and the result would follow with $V = cU_j$ if $\epsilon_j < \epsilon/2$.

Given $p \in M$, suppose for a contradiction that such ϵ_j and δ_j do not exist. Then for all $\epsilon > 0$ sufficiently small and for all $\delta \in (0, \epsilon)$, $c(\overline{B_\delta(p)})$ has infinitely many components which intersect $cB_\epsilon(p)$. Since M is locally compact, suppose without loss that $\overline{B_\epsilon(p)}$ is compact.

Then since M is path-connected, for any $x \in cB_\epsilon(p)$ there is a continuous function $\rho : [0, 1] \to M$ with $\rho(0) = p$ and $\rho(1) = x$, and the function $\mu : [0, 1] \to \mathbb{R}$ defined by $\mu(t) = d(p, \rho(t))$, $t \in [0, 1]$, is continuous with $\mu(0) = 0$ and $\mu(1) \geqslant \epsilon$. Therefore, by the intermediate value theorem there exists $t \in (0, 1)$ with $\mu(t) = \epsilon/2$. Hence any component of $c(\overline{B_{\epsilon/4}(p)})$ which contains a point $x \in cB_\epsilon(p)$ must intersect the compact set $S_{\epsilon/2}(p) := \{y \in M : d(p, y) = \epsilon/2\}$.

Since by assumption $c(\overline{B_{\epsilon/4}(p)})$ has infinitely many components that intersect $cB_\epsilon(p)$, it follows that infinitely many components of $c(\overline{B_{\epsilon/4}(p)})$ intersect $S_{\epsilon/2}(p)$. Suppose $\{q_k : k \in \mathbb{N}\} \subset S_{\epsilon/2}(p)$ is a sequence of points belonging to different components of $c(\overline{B_{\epsilon/4}(p)})$ and, by compactness suppose that $q_k \to q \in S_{\epsilon/2}(p)$. Then $B_{\epsilon/4}(q)$ is a subset of $B_\epsilon(p) \setminus \overline{B_{\epsilon/4}(p)}$ which intersects infinitely many components of $c(\overline{B_{\epsilon/4}(p)})$.

However, M is weakly locally connected at q and for every $\eta > 0$ there is a connected neighbourhood $W \subset B_\eta(q)$ of q. But this is false since $B_{\epsilon/4}(q)$ has infinitely many components and q is the limit of a sequence q_k of points from distinct components.   □

Remark For V as in the statement, V° is an open neighbourhood of p such that $c(V^\circ) = \overline{cV}$ has finitely many components (each being a finite union of the closures of some of the finitely many components of cV).   □

As noted at the beginning of Sect. 3.1, in a metric space M

$$\overline{B_\epsilon(x)} \setminus B_\epsilon(x) = \partial B_\epsilon(x) \subset S_\epsilon(x) = \{y \in M : d(x, y) = \epsilon\},$$

but equality does not always hold. The following result is due to Fraser [19], see also [36].

Theorem 4.26 *For $x \in M$ and $\epsilon > 0$, suppose $\overline{B_\epsilon(x)}$ is compact and for all $\delta \in (0, \epsilon]$*

$$\partial B_\delta(x) = S_\delta(x), \text{ equivalently } \overline{B_\delta(x)} = \{y \in M : d(x, y) \leqslant \delta\}.$$

Then $\overline{B_\delta(x)}$, $\delta \in (0, \epsilon]$, is connected, and M is weakly locally connected at x.

Proof If $\overline{B_\epsilon(x)}$ is not connected, for non-empty closed sets U and V,

$$\overline{B_\epsilon(x)} = U \cup V \text{ where } x \in U \text{ and } U \cap V = \emptyset.$$

So by compactness there is $v \in V$ with $d(x, v) = \text{dist}(x, V) \in (0, \epsilon]$ and so

$$v \in S_{\text{dist}(x,V)}(x) = \partial B_{\text{dist}(x,V)}(x), \text{ by hypothesis.}$$

But $U \cap V = \emptyset$, $B_{\text{dist}(x,V)}(x) \subset U$ and $\overline{B_{\text{dist}(x,V)}(x)} \subset U$, which is a contradiction. This proves that $\overline{B_\epsilon(x)}$, which is a neighbourhood of x, is connected. By the same argument $\overline{B_\delta(x)}$ is connected for all $\delta \in (0, \epsilon]$, and so M is weakly locally connected at x.   □

Corollary 4.27 *If M is locally compact and for all $x \in M$ there exists $\epsilon_x > 0$ with $\partial B_\delta(x) = S_\delta(x)$ for all $\delta \in (0, \epsilon_x]$, then M is locally connected.*

Proof For each $x \in M$, by local compactness $\epsilon_x > 0$ can be chosen so that $\overline{B_\epsilon(x)}$ is compact for all $\epsilon \in (0, \epsilon_x]$, and hence $\overline{B_\epsilon(x)}$ is a connected neighbourhood of x by Theorem 4.26. Therefore M is weakly locally connected, and hence M is locally connected by Lemma 4.23.   □

Chapter 5
Congestion Points

This chapter explores how the failure of a metric space to be weakly locally connected at a point has implications for global bifurcation theory.

Definition 5.1 A point $x \in M$ is a congestion point if M is not weakly locally connected at x. □

Here the abbreviation "congestion point" is adopted to reflect the complex behaviour, see Corollary 5.3 and Theorem 5.8, of generalised continua in neighbourhoods of points at which they are not weakly locally connected. The following results are in the spirit of [49, Ch. I, (13.1), p.19] and Remark 3.51.

Theorem 5.2 *Suppose M is connected and $x \in M$ is such that for all $\epsilon > 0$ there is a neighbourhood U of x and a set S with*

$$\partial U \subset S, \ \mathrm{diam}\,(S \cup U) < \epsilon \ \text{and S has finitely many components.}$$

Then M is weakly locally connected at x.

Proof If $M = \{x\}$, the theorem is obvious, and it can be assumed that $\mathrm{diam}\,(M) \in (0, \infty]$. Let $x \in M$, $\epsilon \in (0, \mathrm{diam}\,(M))$, and let U and S be as in the statement. Then $U \neq M$ and, by Lemma 3.50, $S \cup U$ has finitely many components, $V_1, \cdots, V_m$, where $x \in V_1$ and $x \notin \cup_2^m V_j$. Since $V_j = \overline{V_j} \cap (S \cup U)$ by Remark 3.49 and (3.7a), it follows that $x \notin \overline{V_j}$, $2 \leqslant j \leqslant m$ and $\delta = \mathrm{dist}\,(x, \cup_{j=2}^m V_j) > 0$. Since U is a neighbourhood of x, $B_\eta(x) \subset U$ for some $\eta \in (0, \delta/2)$. Hence $B_\eta(x) \subset V_1$ and so V_1 is a connected neighbourhood of x with $\mathrm{diam}\,(V_1) < \epsilon$. Since $\epsilon > 0$ was arbitrary, M is weakly locally connected at x by Definition 4.21. □

Corollary 5.3 *If x is a congestion point of a connected space M there exists $\epsilon > 0$ such that the boundary of every neighbourhood U of x with $\mathrm{diam}\,(U) < \epsilon$, has infinitely many components.*

© The Author(s), under exclusive license to Springer Nature Switzerland AG 2025 45
B. Buffoni, J. Toland, *Connected Sets in Global Bifurcation Theory*, SpringerBriefs in Mathematics, https://doi.org/10.1007/978-3-031-87051-4_5

Proof If this is false there is a sequence of neighbourhoods $\{U_k\}$ of a congestion point x with diam$(U_k) \to 0$ as $k \to \infty$ and ∂U_k has finitely many components. However, by Theorem 5.2 with $S = \partial U_k$, M is weakly locally connected at x, which contradicts Definition 5.1. $\square$

Corollary 5.4 *Suppose M is connected, and for $x \in M$ and any $\epsilon > 0$ there is a neighbourhood of x with diam$(U) < \epsilon$ such that ∂U is compact and weakly locally connected. Then M is weakly locally connected at x.*

Proof By Lemma 4.23 ∂U is locally connected, and all its components are open in ∂U by Lemma 4.7. Hence, since ∂U is compact, it has finitely many components, and M is weakly locally connected at x by Theorem 5.2. $\square$

5.1   Congestion Points in Metric Subspaces of Banach Spaces

This section leads to Theorem 5.15, which is a criterion for a connected set of solutions of Eq. (1.1) to be path-connected.

Theorem 5.5 *Suppose A is connected in a Banach space $(B, \| \cdot \|)$ and for $a \in A$ there is a decomposition $B = Y \oplus Z$ (direct sum of closed subspaces) and $0 < \gamma_n \searrow 0$ and $0 < s_n \searrow 0$ such that the set*

$$S_n = \left\{ a + y + z \in A : \max\left\{ \frac{\|y\|}{s_n}, \frac{\|z\|}{\gamma_n} \right\} = 1, \; y \in Y, \; z \in Z \right\}$$

has finitely many components for each $n \in \mathbb{N}$.

Then A is weakly locally connected at a.

Remark The hypothesis allows the possibility that $Y = B$ and $Z = \{0\}$. As $B = Y \oplus Z$ with Y and Z closed in the Banach space B, the norms on B defined by $y + z \to \|y + z\|$ and $y + z \to \|y\| + \|z\|$ for $(y, z) \in Y \times Z$ generate the same topology on B; see Theorem 2.10 in [5]. $\square$

Proof By Theorem 5.2 it suffices to show that for all $\epsilon > 0$ there is a neighbourhood U of a in A with diam$(U) < \epsilon$ and $\partial_A U \subset S \subset A$, where S is either empty or S has finitely many components. So for $n \in \mathbb{N}$ let

$$U_n = \left\{ a + y + z \in A : \max\left\{ \frac{\|y\|}{s_n}, \frac{\|z\|}{\gamma_n} \right\} < 1, \; y \in Y, \; z \in Z \right\}.$$

Then U_n is a neighbourhood of a in A with diam$(U_n) \to 0$ as $n \to \infty$ and, by (3.7d), $\partial_A U_n \subset S_n$ where by hypothesis S_n has finitely many components. This completes the proof. $\square$

Corollary 5.6 *Suppose A is a generalised continuum in a Banach space $(B, \| \cdot \|)$ and for $a \in A$ there is a decomposition $B = Y \oplus Z$ and $0 < \gamma_n \searrow 0$ and $0 < s_n \searrow 0$ such that, for all $n \in \mathbb{N}$ and all $s \in (0, s_n]$,*

$$\{a + y + z \in A : \|y\| = s, \|z\| \leqslant \gamma_n, \ y \in Y, \ z \in Z\}$$

has finitely many components

$$\text{and} \ \{a + z \in A : \|z\| = \gamma_n, \ z \in Z\} = \emptyset.$$

Then A is weakly locally connected at a.

Remark If the aim is to show that the set A is weakly locally connected, the decompositions may be different for different $a \in A$. □

Proof By the previous theorem, it suffices to check that for each $n \in \mathbb{N}$ sufficiently large, there exists $s \in (0, s_n]$ such that

$$\{a + y + z \in A : \|y\| \leqslant s, \|z\| = \gamma_n, \ y \in Y, \ z \in Z\} = \emptyset.$$

Otherwise, for some arbitrarily large n there exist sequences $\{y_k\} \subset Y$ and $\{z_k\} \subset Z$, with

$$a + y_k + z_k \in A, \ \|y_k\| \leqslant s_n/k, \ \|z_k\| = \gamma_n, \quad k \in \mathbb{N}.$$

Since A is locally compact, for some fixed n sufficiently large the sequence $\{a + y_k + z_k\}$ has a convergent subsequence, leading to the contradiction $\{a + z \in A : \|z\| = \gamma_n, \ z \in Z\} \neq \emptyset$. □

Corollary 5.7 *Suppose A in Theorem 5.5 is a generalised continuum and, for some $a \in A$ there exists $\gamma_0 > 0$ such that for all $s \in [0, \gamma_0]$ the set*

$$\{a + y + z \in A : \|y\| = s, \|z\| \leqslant \gamma_0, \ y \in Y, \ z \in Z\} \ \text{is finite.} \tag{5.1}$$

Then A is weakly locally connected at a.

Proof Note that if the assumption holds for $\gamma_0 > 0$, it holds for any smaller value of γ_0. Moreover the choice $s = 0$ gives that

$$\{a + z \in A : \|z\| \leqslant \gamma_0, \ z \in Z\}$$

is finite. Therefore, by decreasing the size of γ_0, it can further be assumed that

$$\{a + z \in A : \|z\| \leqslant \gamma_0, \ z \in Z\} = \{a\}.$$

To complete the proof choose two sequences $0 < \gamma_n \searrow 0$ and $0 < s_n \searrow 0$ such that, for all $n \in \mathbb{N}$, $\{\gamma_n, s_n\} \subset (0, \gamma_0]$, and apply Corollary 5.6. □

5.2   Congestion Points in Generalised Continua

The next result describes the complex behaviour of a generalised continuum in arbitrarily small neighbourhoods of a congestion point.

Theorem 5.8 *When x is a congestion point of a generalised continuum M there exists $\epsilon_0 > 0$ such that for all $\epsilon \in (0, \epsilon_0)$ there is a non-degenerate continuum $K \subset \overline{B_\epsilon}(x)$ such that $x \in K$, $K \cap \partial B_\epsilon(x) \neq \emptyset$ and every point of $K \cap B_\epsilon(x)$ is a congestion point.*

Moreover, for a sequence $\{K_j\}$ of distinct components of $B_\epsilon(x)$,

$$x \notin \overline{K_j}, \quad \overline{K_j} \cap \partial B_\epsilon(x) \neq \emptyset \text{ for all } j \in \mathbb{N}, \text{ and } \lim K_j = K$$

in the sense of Definition 3.42.

Since it is not claimed that $K \cap B_\epsilon(x)$ is connected, let $K(x)$ be the component of $K \cap B_\epsilon(x)$ which contains x. Then $K(x) \cap K_j = \emptyset$ for all $j \in \mathbb{N}$, $K(x)$ has a limit point in $\partial B_\epsilon(x)$ and $K(x)$ is a non-degenerate connected set of congestion points of M.

Remark Whyburn [49, p. 18] refers to K as a continuum of convergence and Nadler refers to it as a convergence continuum [37, § 5.10, p. 75]. See also [23, Thm. 3-12, p. 114] and [51, Thm. 2.1, p. 102].   □

Proof Since M is locally compact, by Definition 3.2 there exists $\epsilon_0 > 0$ such that for all $\epsilon \in (0, \epsilon_0)$ $\overline{B_\epsilon}(x) \neq M$ is compact. Then, since M is not weakly locally connected at x, by Definition 4.21 ϵ_0 can be chosen such that, for $\epsilon \in (0, \epsilon_0)$ and any connected set U with $x \in U \subset B_\epsilon(x)$, $B_\delta(x)$ for any $\delta > 0$ is not a subset of U. Now recall that M is connected and let $\epsilon \in (0, \epsilon_0)$.

Then for all $\delta_1 \in (0, \epsilon)$ the component U of $B_\epsilon(x)$ that contains x is such that $B_{\delta_1}(x) \not\subset U$. Hence there exists $x_1 \in B_{\delta_1}(x)$ such that $x \notin K_1$ where K_1 is the component of $B_\epsilon(x)$ which contains x_1. Because the closure of K_1 in $B_\epsilon(x)$ is $\overline{K_1} \cap B_\epsilon(x)$ and K_1 is closed in $B_\epsilon(x)$ (Remark 3.49), it follows that $x \notin \overline{K_1}$, and $\overline{K_1} \cap \partial B_\epsilon(x) \neq \emptyset$, by Corollary 3.55 (ii) with $A = M$ and $G = B_\epsilon(x)$.

Let $\delta_2 = \frac{1}{2}\text{dist}(K_1, x) \in (0, \delta_1/2)$. By the same argument there exists $x_2 \in B_{\delta_2}(x)$ such that $x \notin K_2$ where K_2 is the component of $B_\epsilon(x)$ which contains x_2 and $\overline{K_2} \cap \partial B_\epsilon(x) \neq \emptyset$. Moreover, $K_1 \cap K_2 = \emptyset$ since $x_2 \notin K_1$ and K_2 is a component of $B_\epsilon(x)$. Now let $\delta_3 = \frac{1}{2}\text{dist}(\overline{K_2}, x)$ and proceed by induction to obtain sequences $\delta_j \searrow 0$ in $\mathbb{R}$, $x_j \in B_{\delta_j}(x)$, and components K_j of $B_\epsilon(x)$ such that

$$x_j \in K_j, \quad x \notin \overline{K_j}, \quad \overline{K_j} \cap \partial B_\epsilon(x) \neq \emptyset, \quad K_i \cap K_j = \emptyset, i \neq j \in \mathbb{N}.$$

Then by Remark 3.45 there is no loss in assuming that $K = \lim K_j$ exists where $K \subset \overline{B_\epsilon}(x)$ is a non-empty, closed, connected set. Moreover, since $\overline{B_\epsilon}(x)$ is compact and $\overline{K_j} \cap \partial B_\epsilon(x) \neq \emptyset$ for all j, it follows that $K \cap \partial B_\epsilon(x) \neq \emptyset$, and $x \in K$ since $x_j \to x$ where $x_j \in K_j$.

To show that every $q \in K \cap B_\epsilon(x)$ is a congestion point let $\delta = \epsilon - d(x, q) > 0$. Since $B_\delta(q) \subset B_\epsilon(x)$ and $q \in K = \lim K_j$, for a subsequence if necessary there exists $q_j \in K_j \cap B_\delta(q)$ with $q_j \to q$. Therefore, every neighbourhood V of q with diam $(V) < \delta/2$ has infinitely many components, and hence all points q in $K \cap B_\epsilon(x)$ are congestion points of M.

Now let $K(x)$ denote the component of $K \cap B_\epsilon(x)$ that contains x. Then since $x \notin K_j$ it follows that $K_j \cap K(x) = \emptyset$ for all $j \in \mathbb{N}$. Finally, by Corollary 3.55 (ii) with $A = K$ and $G = B_\epsilon(x)$, $K(x)$ has a limit point in $\partial B_\epsilon(x)$. Thus $K(x)$ is a non-degenerate, connected set of congestion points of M, which completes the proof. □

Remark 5.9 If a metric space M fails to have connected neighbourhoods of x of arbitrarily small diameter it is immediate that x is a congestion point of M. Consequently, x is a congestion point of a generalised continuum if and only if the conclusion of Theorem 5.8 holds. □

Definition 5.10 Let $\mathscr{N}(M)$ denote the set of all congestion points of M and let $N[x]$ denote the component of $\mathscr{N}(M)$ which contains $x \in \mathscr{N}(M)$. If M is a generalised continuum $N[x]$ is non-degenerate by Theorem 5.8. □

Remark 5.11 That congestion points of M may not be congestion points of $\mathscr{N}(M)$ is evident from the topologist's sine curve $\overline{T}$ in Example 4.5. There $\overline{T}$ is locally connected everywhere except at its congestion points $\mathscr{N}(\overline{T}) = \{0\} \times [-1, 1]$ which is a straight-line segment. Hence the metric subspace $\mathscr{N}(\overline{T})$ is locally connected, but $\overline{T}$ is not locally connected at points of $\mathscr{N}(\overline{T})$.

The continuum $T^* := \overline{T} \cup \big([0, 1/\pi] \times \{0\}\big)$ is path-connected with $\mathscr{N}(T^*) = \big(\{0\} \times [-1, 1]\big) \setminus \{(0, 0)\}$ which is the union of two non-degenerate components zero distance apart. □

5.3 Partitioning a Generalised Continuum

The following hypotheses are motivated by questions about connected sets of solutions of equations such as (1.1) in global bifurcation theory.

Theorem 5.12 *When M is a generalised continuum such that $M \neq \overline{\mathscr{N}(M)}$, the components Q of $M \setminus \mathscr{N}(M)$ are path-connected, and for every $p \in Q$ and $\epsilon > 0$ there is a closed neighbourhood V of p in Q with diam $(V) < \epsilon$ for which $Q \setminus V$ has finitely many components. There are two cases:*

(A) $\mathscr{N}(M) = \emptyset$, in which case $Q = M$.

(B) $\mathscr{N}(M) \neq \emptyset$ in which case $\emptyset \neq \partial Q \subset \partial \mathscr{N}(M)$, and for $z \in \partial Q$ either

> *(i) $z \in \mathscr{N}(M)$ and $Q \cup N[z]$ is connected, or*
> *(ii) $z \notin \mathscr{N}(M)$ and for any $\epsilon > 0$ there is a connected closed neighbourhood V of z with diam $(V) < \epsilon$ and $y \in V \cap \mathscr{N}(M)$ such that $Q \cup V \cup N[y] \subset M$ is connected.*

Proof The proof deals with the cases in turn.

(A) If $\mathcal{N}(M) = \emptyset$, M has no congestion points and $Q = M$ is path-connected by Theorem 4.12 and Lemma 4.23. Moreover, by Theorem 4.25, for any $p \in Q$ and $\epsilon > 0$ there exists a closed neighbourhood V of p in Q with diam $(V) < \epsilon$ for which $Q \setminus V$ has finitely many components.

(B) If $\mathcal{N}(M) \neq \emptyset$ and $M \setminus \mathcal{N}(M) \neq \emptyset$, M is weakly locally connected at every point of the open set $M \setminus \overline{\mathcal{N}(M)}$ and hence $M \setminus \overline{\mathcal{N}(M)}$ is weakly locally connected. Therefore, by Lemma 4.23, $M \setminus \overline{\mathcal{N}(M)}$ is locally connected and, by Lemma 4.7, the component Q of $M \setminus \overline{\mathcal{N}(M)}$ is open.

Now since $\mathcal{N}(M) \neq \emptyset$ it follows that $\partial Q \neq \emptyset$ because otherwise Q is open and closed in M, and this is false since M is connected and $M \neq Q \neq \emptyset$. If $z \in \partial Q \setminus \overline{\mathcal{N}(M)}$ then $z \notin Q$ because Q is open, $Q \cup \{z\}$ is connected by Lemma 3.25, and $Q \cup \{z\} \subset M \setminus \overline{\mathcal{N}(M)}$, which contradicts the fact that Q is a component of $M \setminus \overline{\mathcal{N}(M)}$. This proves that $\partial Q \subset \overline{\mathcal{N}(M)}$. If $z \in (\mathcal{N}(M))^\circ$, for $\delta > 0$ sufficiently small, $B_\delta(z) \subset \mathcal{N}(M) \subset M \setminus Q$ which implies that $z \notin \partial Q$. Hence $\emptyset \neq \partial Q \subset \partial\mathcal{N}(M)$ when $\mathcal{N}(M) \neq \emptyset$.

Then by Corollary 4.9 Q is a locally connected generalised continuum, and by Theorem 4.12 it is path-connected. Moreover, by Theorem 4.25, for any $p \in Q$ and $\epsilon > 0$ there exists a closed neighbourhood V of p in Q with diam $(V) < \epsilon$ for which $Q \setminus V$ has finitely many components.

Now for $z \in \mathcal{N}(M)$ let $N[z]$ be the component of $\mathcal{N}(M)$ in Definition 5.10.

(B)(i) Suppose $z \in \mathcal{N}(M) \cap \partial Q$. Then $Q \cup N[z]$ is connected because $Q \cup \{z\}$ and $N[z]$ are connected and z belongs to both.

(B)(ii) Suppose $z \in (\partial Q) \setminus \mathcal{N}(M)$ and let $\epsilon > 0$. Since M is weakly locally connected at z, there exists a closed connected neighbourhood V of z in M with diam $(V) < \epsilon$ and $V \cap Q \neq \emptyset \neq V \cap \mathcal{N}(M)$ because $z \in (\partial\mathcal{N}(M)) \cap \partial Q$. Hence $V \cup Q \cup N[y]$ is connected when $y \in V \cap \mathcal{N}(M)$.

This completes the proof. $\qquad\qquad\qquad\qquad\qquad\qquad\qquad\qquad\qquad\qquad\qquad\qquad\square$

Example 5.13 (Theorem 5.12 (B)(i): $z \in \mathcal{N}[M] \cap \partial Q$) In Theorem 5.12 (B) let $M = \overline{T}$, the topologist's sine curve in Example 4.5, and let $p = (1/\pi, 0)$. Then

$$M = \{(t, \sin(1/t)) : t > 0\} \cup (\{0\} \times [-1, 1]) \subset \mathbb{R}^2,$$

$$p \in Q = \{(t, \sin(1/t)) : t > 0\}, \quad \mathcal{N}(M) = \{0\} \times [-1, 1] = \partial Q = \partial\mathcal{N}(M).$$

Since in Definition 5.10, $N[z] = \mathcal{N}(M)$ for all $z \in \partial Q \cap \partial\mathcal{N}(M)$, this is an example of Theorem 5.12 (B)(i) when $\mathcal{N}(M) \cap \partial Q \neq \emptyset$. $\qquad\qquad\qquad\square$

The following examples which illustrate Theorem 5.12 (B)(ii) involve the infinite broom referred to in Remark 4.24.

The Infinite Broom

Let $\mathscr{B} = \bigcup_{k=0}^{\infty} B_k \subset \mathbb{R}^2$ where $B_0 = \{(0, 0)\} \subset \mathbb{R}^2$ and

$$B_k = \left(\left[\frac{1}{k+1}, \frac{1}{k}\right] \times \{0\}\right) \bigcup \left\{\left(t, \frac{1-kt}{kn}\right) : t \in \left[\frac{1}{k+1}, \frac{1}{k}\right], n \in \mathbb{N}\right\}, \quad k \in \mathbb{N},$$

Fig. 5.1 The infinite broom,
see e.g. [44, p. 139]

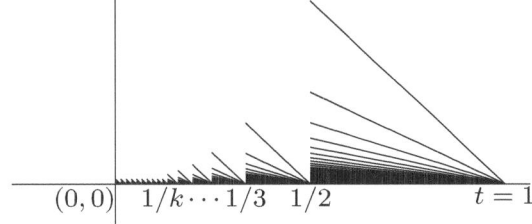

$$(0,0) \quad 1/k \cdots 1/3 \quad 1/2 \qquad\qquad t=1$$

as illustrated in Fig. 5.1 in which each B_k, $k \in \mathbb{N}$, is a re-scaling of B_1. Clearly, $\mathscr{B}$ is locally connected at every point of $\mathscr{B} \setminus ([0,1] \times \{0\})$, $\mathscr{B}$ is not weakly locally connected at any point of $(0,1) \times \{0\} \subset \mathscr{B}$ and $\mathscr{B}$ is locally connected at $(1,0)$. However $\mathscr{B}$ is not locally connected, but it is weakly locally connected, at $(0,0)$.

To see that $\mathscr{B}$ is not locally connected at $(0,0)$, let C be a connected open neighbourhood of $(0,0)$ in $\mathscr{B}$. Then $(1/(k+1),0) \in C$ for some $k \in \mathbb{N}$ and, since C is open and connected, it follows that $(1/k,0) \in C$. From the same argument it follows that $(1/(k-1),0) \in C$ and after k steps it emerges that $(1,0) \in C$ when C is a connected open neighbourhood of $(0,0)$. This shows that $\mathscr{B}$ is not locally connected at $(0,0)$. However, since $\left\{B_0 \cup \left(\cup_{k \geqslant n} B_k\right)\right\}_{n \geqslant 1}$ is a family of connected neighbourhoods of $(0,0)$ in $\mathscr{B}$ with diameters tending to 0 as $n \to \infty$, $\mathscr{B}$ is weakly locally connected at $(0,0)$. □

Examples 5.14 (Theorem 5.12 (B)(ii): $z \in \partial Q$, $z \notin \mathcal{N}[M]$)

(I) In Fig. 5.2, $M = (\mathbb{R} \times \{0\}) \cup \mathscr{B}$, where $\mathscr{B}$ is the infinite broom, and $p = (-1,0)$. Since $\overline{\mathcal{N}(M)} = [0,1] \times \{0\}$ and $(0,0) \notin \mathcal{N}(M)$,

$$p \in Q = (-\infty, 0) \times \{0\}, \quad \partial Q = \{(0,0)\} \subset \partial \mathcal{N}(M) \setminus \mathcal{N}(M),$$

and for any $y \in V \cap \mathcal{N}(M)$, where V is a connected neighbourhood of $(0,0)$,

$$N[y] = \mathcal{N}(M) = (0,1) \times \{0\} \text{ independent of } V,$$

and diam $([N(y)]) \nrightarrow 0$ as diam $(V) \to 0$ in Theorem 5.12 (B)(ii).

(II) In Fig. 5.3 gaps have been introduced between B_k and B_{k+1} in Example (I). To be precise, let $\{\sigma_k\}_{k \geqslant 1}$ with $\sigma_1 = 0$ be an increasing sequence of non-negative numbers that converge to $\sigma \in (0, \infty)$. Then define

$$D_0 = B_0 - (\sigma, 0) = \{(-\sigma, 0)\}, \quad D_k = B_k - (\sigma_k, 0) \ (k \geqslant 1);$$

in other words, for $k \geqslant 1$, D_k is a shift of B_k to the left through σ_k.

Now define a connected, locally compact set M by

$$M = (\mathbb{R} \times \{0\}) \cup \mathscr{D} \text{ where } \mathscr{D} = \bigcup_{k=0}^{\infty} D_k \subset \mathbb{R}^2,$$

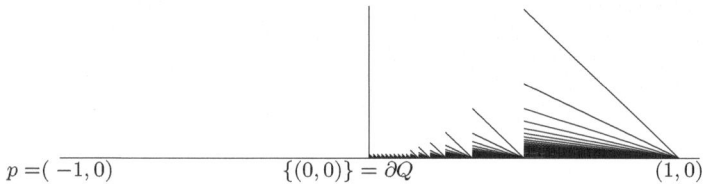

$p =(\overline{-1,0})$ $\{(0,0)\} = \partial Q$ $(1,0)$

Fig. 5.2 diam $([N(y)]) \nrightarrow 0$ as diam $(V) \rightarrow 0$ in Theorem 5.12 (B)(ii)

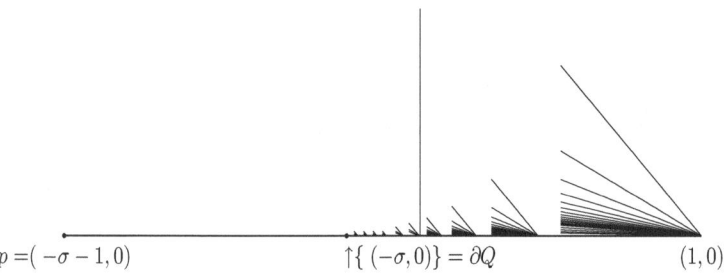

$p =(-\sigma - 1,0)$ $\uparrow\{ (-\sigma,0)\} = \partial Q$ $(1,0)$

Fig. 5.3 diam $([N(y)]) \rightarrow 0$ if diam $(V) \rightarrow 0$ in Theorem 5.12 (B)(ii)

and note that M is locally connected at every point of $M \backslash ([-\sigma, 1] \times \{0\})$. In addition, because of the gaps between the D_ks, M is locally connected at $(-\sigma, 0) \in D_0$ and at the right end-points, $(k^{-1} - \sigma_k, 0), k \in \mathbb{N}$, of the closed segments $D_k \cap (\mathbb{R} \times \{0\})$.

All the other points of $\mathscr{D} \cap (\mathbb{R} \times \{0\})$ are congestion points and $\mathscr{N}(M)$ has infinitely many components, namely the segments $D_k \cap (\mathbb{R} \times \{0\})$, $k \in \mathbb{N}$, with their right end-points removed:

$$\mathscr{N}(M) = \cup_{k \geqslant 1} J_k \text{ where } J_k := \left[\frac{1}{k+1} - \sigma_k, \frac{1}{k} - \sigma_k \right) \times \{0\}, \quad k \geqslant 1.$$

Therefore, if $p = \left(-(\sigma + 1), 0 \right)$ in Theorem 5.12 (B)(ii),

$$Q = (-\infty, -\sigma) \times \{0\}, \quad \partial Q = \{(-\sigma, 0)\} \subset \partial \mathscr{N}(M) \setminus \mathscr{N}(M),$$

and $\{J_k\}$ is a sequence of components of $\mathscr{N}(M)$ with diam $(J_k) \rightarrow 0$ and dist$\left(J_k, (-\sigma, 0)\right) \rightarrow 0$ as $k \rightarrow \infty$. Thus for any connected neighbourhood V of $(-\sigma, 0)$ there exists $y \in V \cap \mathscr{N}(M)$ for which the component $N[y]$ of $\mathscr{N}(M)$ is a subset of V.

In contrast with part (I) in this Example

$$\text{diam } ([N(y)]) \rightarrow 0 \text{ if diam } (V) \rightarrow 0 \text{ in Theorem 5.12 (B)(ii)}. \qquad \square$$

5.4 Global Bifurcation Theory

When the hypotheses of Theorem 1.3 are satisfied, $L : X \to X$ and $R : \mathbb{R} \times X \to X$ are continuous and compact, which implies that any set of solutions of (1.1a) which is closed and bounded in $\mathbb{R} \times X$ is compact. In particular, $\overline{\mathscr{C}}$ in Theorem 1.3 with the metric inherited from $\mathbb{R} \times X$ is a generalised continuum. There follows an account of the relevance to the global bifurcation theory of Eq. (1.1) of Theorem 5.12 when R satisfies the hypothesis of Theorem 1.3 (b)(i) or (ii).

For convenience, let $\mathscr{M}$ denote $\overline{\mathscr{C}}$ with the metric inherited from $\mathbb{R} \times X$, let $B_\epsilon(\lambda, x)$ be the open ball in $\mathbb{R} \times X$ centred at (λ, x), and if (λ, x) is a congestion point of $\mathscr{M}$ let $K(\lambda, x)$ correspond to the connected subset $K(x)$ of M in the last paragraph of Theorem 5.8.

Then when (λ, x) is a congestion point in the generalised continuum $\mathscr{M}$, by Theorem 5.8 there exists, for any arbitrarily small $\epsilon > 0$, a non-degenerate connected subset $K(\lambda, x)$ of $\mathscr{M}$ with the following properties:

(i) $(\lambda, x) \in K(\lambda, x) \subset B_\epsilon(\lambda, x)$;
(ii) every point of $K(\lambda, x)$ is a congestion point of $\mathscr{M}$;
(iii) $K(\lambda, x)$ has a limit point in $\partial_{\mathscr{M}}(B_\epsilon(\lambda, x) \cap \mathscr{M}) \subset \mathscr{M} \cap \partial B_\epsilon(\lambda, x)$;
(iv) $K(\lambda, x) \cap K_j = \emptyset$ and $\overline{K_j} \cap \partial_{\mathscr{M}}(B_\epsilon(\lambda, x) \cap \mathscr{M}) \neq \emptyset$ for a sequence $\{K_j\}$ of distinct components of $B_\epsilon(\lambda, x) \cap \mathscr{M}$;
(v) $K(\lambda, x) \subset \lim K_j$, see Definition 3.42;
(vi) by Corollary 5.3 the boundary in $\mathscr{M}$ of every neighbourhood in $\mathscr{M}$ of (λ, x) with diameter less than ϵ has infinitely many components.

On the other hand, by Corollary 5.7 with $Y = \mathbb{R}$ and $Z = X$, (λ, x) is <u>not</u> a congestion point of $\mathscr{M}$ if

(vii) there exists $\gamma_0 > 0$ such that, for all $s \in [0, \gamma_0]$,

$$\text{the set } \left\{ (\hat{\lambda}, \hat{x}) \in \mathscr{M} : |\lambda - \hat{\lambda}| = s, \ \|x - \hat{x}\| \leqslant \gamma_0 \right\} \text{ is finite.} \quad \square$$

Moreover, by Theorem 1.3 (b) there is a homeomorphism $\gamma : (-1, 1) \to \mathscr{M} \cap U$ with $\gamma(0) = (\lambda_0, 0)$. Hence for any $(\lambda, x) \in \mathscr{M} \cap U$, there is a unique $t \in (-1, 1)$ such that $(\lambda, x) = \gamma(t)$. Therefore for $\epsilon > 0$ there exists $\delta \in (0, 1 - |t|)$ such that, in the metric subspace $\mathscr{M}$, $\{\gamma(s + t) : s \in (-\delta, \delta)\} \subset U \cap \mathscr{M}$ is an open connected neighbourhood of (λ, x) with diameter less than ϵ. This shows that no point of $\mathscr{M} \cap U$ is a congestion point and in particular $(\lambda_0, 0) \in \mathscr{M} \setminus \overline{\mathscr{N}(\mathscr{M})}$.

Therefore, by Theorem 5.12 there are the following possibilities.

(α) $\mathscr{N}(\mathscr{M}) = \emptyset$, $\mathscr{M}$ is path-connected and every $(\lambda, x) \in \mathscr{M}$ has a closed neighbourhood V in $\mathscr{M}$ of arbitrarily small diameter such that $\mathscr{M} \setminus V$ has finitely many components.
(β) $\mathscr{N}(\mathscr{M}) \neq \emptyset$ and the component Q of $\mathscr{M} \setminus \overline{\mathscr{N}(\mathscr{M})}$ which contains $(\lambda_0, 0)$ is open in $\mathscr{M}$, path-connected, and every $(\lambda, x) \in Q$ has a closed neighbourhood

V in Q of arbitrarily small diameter such that $Q \setminus V$ has finitely many components.

Furthermore, in (β) the boundary of Q in $\mathscr{M}$ is a non-empty subset of the boundary of $\mathscr{N}(\mathscr{M})$ in $\mathscr{M}$ and for any $(\lambda, x) \in \partial_{\mathscr{M}} Q$, either

(β)(i) $(\lambda, x) \in \mathscr{N}(\mathscr{M})$, in which case $Q \cup N[\lambda, x]$ is connected where $N[\lambda, x]$ is the non-degenerate component of $\mathscr{N}(\mathscr{M})$ which contains (λ, x); or

(β)(ii) $(\lambda, x) \notin \mathscr{N}(\mathscr{M})$ and for any $\epsilon > 0$ there exists a closed connected neighbourhood V of (λ, x) with $\mathrm{diam}\,(V) < \epsilon$ and $(\mu, y) \in V \cap \mathscr{N}(\mathscr{M})$ such that $Q \cup V \cup N[\mu, y] \subset \mathscr{M}$ is connected.

Then there are further possibilities depending on whether

$$\liminf_{\epsilon \to 0} \left(\mathrm{diam}\,(N[\mu, y]) \right) \text{ is zero or not,}$$

when $(\mu, y) \in \mathscr{N}(\mathscr{M}) \cap V$ and $\mathrm{diam}\,(V) < \epsilon \to 0$.

For $(\lambda, x) \in (\partial_{\mathscr{M}} Q) \setminus \mathscr{N}(\mathscr{M})$ and connected neighbourhoods V_k of (λ, x) with $\mathrm{diam}\,(V_k) \to 0$, there may exist $(\hat{\mu}_k, \hat{y}_k)$ and $(\tilde{\mu}_k, \tilde{y}_k)$ in $\mathscr{N}(\mathscr{M}) \cap V_k$ such that

$$\liminf_{k \to \infty} \left(\mathrm{diam}\,(N[\hat{\mu}_k, \hat{y}_k]) \right) = 0 \text{ and } \liminf_{k \to \infty} \left(\mathrm{diam}\, N[\tilde{\mu}_k, \tilde{y}_k]) \right) > 0.$$

So these possibilities are not mutually exclusive. As noted in Remark 5.11, $\mathscr{M}$ might be path-connected even when $\mathscr{M}$ has congestion points.

When R is real-analytic, Dancer's theory [7] outlined in Sect. 1.4 yields a significantly stronger version of (α). However, when more is known a priori about the solution set $\overline{\mathscr{C}} = \mathscr{M}$, Theorem 5.5 yields the following result on path-connectedness for problem (1.1) without real-analyticity.

Theorem 5.15 *In addition to the hypotheses of Theorem 1.3 suppose, for every fixed $\lambda \in \mathbb{R}$, that all solutions x of (1.1a) are isolated in X. Then $\overline{\mathscr{C}} = \mathscr{M}$ is path-connected and every $(\lambda, x) \in \overline{\mathscr{C}}$ has a closed neighbourhood V of arbitrarily small diameter for which $\overline{\mathscr{C}} \setminus V$ has finitely many components.*

Proof In the notation of Sect. 5.1, let $B = \mathbb{R} \times X$, $Y = \mathbb{R}$, $Z = X$, and $A = \overline{\mathscr{C}}$ where $\mathscr{C}$ is as in Theorem 1.3. Then, for a fixed λ and any closed bounded set $F \subset X$, the set $\{(\lambda, x) \in \overline{\mathscr{C}} : x \in F\}$ is a compact set of isolated points. Thus the hypotheses of Corollary 5.7 are satisfied and hence $\overline{\mathscr{C}}$ is a generalised Peano continuum. The result then follows by Theorem 4.25. $\qquad \square$

Remark 5.16 Example B in Sects. 7.1 and 7.4 is an illustration of (1.1) in which $X = \mathbb{R}$, $R : \mathbb{R} \times X \to X$ is infinitely differentiable, $\lambda_0 = 1$ is a simple eigenvalue and, in the notation of Theorem 1.3, $\overline{\mathscr{C}} = \mathscr{C}^+ \cup \mathscr{L} \cup \mathscr{C}^-$ where $\mathscr{L} = \{1\} \times (-\frac{1}{2}, \frac{1}{2})$, $\mathscr{C}^{\pm}$ are closed generalised continua that contain no non-trivial paths, $\mathscr{C}^+$, $\mathscr{L}$ and $\mathscr{C}^-$ are mutually disjoint, and $(1, \pm\frac{1}{2}) \in \mathscr{C}^{\pm}$.

Then with $p = (1, 0)$ in Theorem 5.12, by the construction of Example B in Sect. 7.1, $\partial Q = \{(1, \frac{1}{2}), (1, -\frac{1}{2})\} \subset \partial Q \cap \mathcal{N}(\overline{\mathscr{C}})$, and this is an example of Theorem 5.12 (b)(i) and of alternative (β)(i) above.

More generally, constructions such as for Examples 5.13 and 5.14, when combined with the method of Sects. 7.3 and 7.4, lead to examples in which R is infinitely differentiable and the possibilities in (β)(ii) are realised; in Sect. 7.5 two distinct possibilities in (β)(ii) arise in the same example.

In conclusion, under the hypotheses of Theorem 1.3, the generalised continuum $\overline{\mathscr{C}}$ which satisfies the conclusion of the theorem also satisfies one of (α), (β)(i) and (β)(ii), and all of these can occur, even when R is smooth.   □

This theory is relevant to problems of the form $x = G(\lambda, x)$ other than (1.1), for example [16, Thm. 1] and [40, eqn. (3.1)] treat problems where $G : \mathbb{R} \times X \to X$ is compact but there is no line of trivial solutions. Note also that G need not be compact, as in [47] which deals with k-set contractions.

5.5 Extending Real-Analytic Theory

In this section the fact that domains of real-analytic functions are always open, Remark 5.20, and the fact that components of congestion points of generalised continua are always non-degenerate, Theorem 5.8, will be merged to yield a *path-connected* global bifurcation theory when the data is not everywhere real-analytic. The significance of these properties was first recognised by Dancer in his seminal work on real-analytic global bifurcation theory [15, § 3], and what follows is a systematic account of their consequences.

Definition 5.17 A metric space M is locally path-connected at x if every neighbourhood of x contains an open path-connected neighbourhood of x. If M is locally path-connected at all points, M is locally path-connected.   □

If M is locally path-connected at x it is locally connected at x, and hence x is not a congestion point of M. The following result is obvious from the definitions.

Lemma 5.18 *If M is weakly locally connected at x and C is the component of M containing x, then C is weakly locally connected at x and x belongs to the interior of C. Hence if x is a congestion point of C, x is a congestion point of M.*   □

Let E and F be Banach spaces and let $U \subset E$ be open.

Definition 5.19 When $\emptyset \neq U \subset E$ is open, a continuous function $G : U \to F$ is said to be real-analytic at $x \in U$ if there is a ball $B_r(x) \subset U$ centred at x on which G is infinitely differentiable and $G(y)$, for all $y \in B_r(x)$, equals the sum of the Taylor series of G at x, the convergence being uniform in $y \in B_r(x)$.   □

Remark 5.20 This definition means that being infinitely differentiable on an open neighbourhood of the point is not enough, see Lemma 7.2. For a discussion of four different but equivalent definitions of real-analyticity at a point, see [18, § 3.1.24, p. 237].

It follows that G is real-analytic at every point of $B_r(x)$, and hence that the set $\{x \in U : G \text{ is real-analytic at } x\}$ is open.   □

Finite Dimensions

In Definition 5.19 let $E = \mathbb{R}^n$, $F = \mathbb{R}^m$ and replace G with g.

Lemma 5.21 *When $\emptyset \neq U \subset \mathbb{R}^n$ is open and $g : U \to \mathbb{R}^m$, $m, n \geqslant 1$, is real-analytic on U, the set $\Omega = \{x \in U : g(x) = 0 \in \mathbb{R}^m\}$ if non-empty is a real-analytic variety in $\mathbb{R}^n$, Ω is locally path-connected, and therefore Ω has no congestion points.*

Proof See [15, §1.1], and specifically Lemma 2.   □

Definition 5.22 A set in a metric space is totally disconnected if it is empty or if all its components are singletons.   □

Remark Any countable set in a metric space is totally disconnected by Corollary 3.24, but the Cantor ternary set [23, p. 97] in the interval $[0, 1]$ is a compact, totally disconnected set which is not countable.   □

Theorem 5.23 *For a continuous function $g : \mathbb{R}^n \to \mathbb{R}^m$ let*

$$Z = \left\{ x \in \mathbb{R}^n : g(x) = 0 \right\} \neq \emptyset,$$

and for a component C of Z suppose

$$N_C = \left\{ x \in C \text{ at which } g \text{ is not real-analytic} \right\} \text{ is totally disconnected.}$$

Then C is path-connected.

Proof Since g is continuous, Z is closed, and by Remark 3.49 any component C of Z is closed in Z, and hence in $\mathbb{R}^n$. Therefore, since $C \subset \mathbb{R}^n$, C is locally compact. If C is not path-connected, by Lemma 4.23 and Theorem 4.12 it has a congestion point and, by Lemmas 5.18 and 5.21, $\mathcal{N}(C) \subset \mathcal{N}(Z) \cap C \subset N_C$. But this is impossible if N_C is totally disconnected because all the components of $\mathcal{N}(C)$ are non-degenerate by Theorem 5.8. Hence C is path-connected when N_C is totally disconnected.   □

Remark 5.24 The conclusion of Theorem 5.23 may be false if $C \subset Z$ is a continuum but not a component of Z. For example, when $g : \mathbb{R}^2 \to \mathbb{R}$ is the zero function which is real-analytic on $\mathbb{R}^2$, $Z = \mathbb{R}^2$ and Knaster's Theorem 6.11 ensures that Z contains a non-degenerate continuum which contains no non-trivial paths.   □

Fig. 5.4 The function g is real-analytic at points of Z except at $(0, 1)$ and $(0, -1)$, and Z, which is unbounded and path-connected, is not a curve

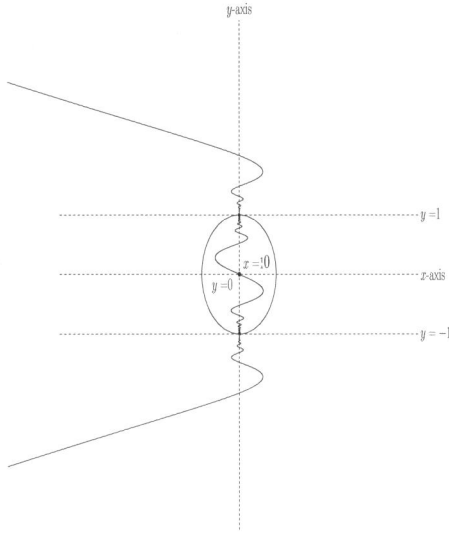

Example For $n = 2$ and $m = 1$ let $g(x, y) = f(x, y)h(x, y)$, $(x, y) \in \mathbb{R}^2$, where

$$f(x, y) = \begin{cases} x - (1 - y^2) \sin\left(\frac{y}{1-y^2}\right), & |y| \neq 1 \\ x, & |y| = 1 \end{cases},$$

$$h(x, y) = 16x^2 + 9y^2 - 9.$$

Then g is real-analytic on $\mathbb{R}^2$ except on the two lines $y = \pm 1$, $x \in \mathbb{R}$, and the zero set Z of g intersects these lines at only two points. Therefore by Theorem 5.23 every components $\mathcal{C}$ of Z is path-connected and in particular, Z is path-connected. However, the unique continuation by local real-analytical re-parametrization of solution curves that pass through a point, which is an important feature of Dancer's theory [7, Thm. 9.1.1], is not meaningful at points of $\mathcal{N}_Z$, see Fig. 5.4.

Infinite Dimensions

Lemma 5.25 *Suppose E is a Banach space, $Z \subset E$ is locally compact and $W \subset Z$ is a non-empty open set in Z. Then if $\mathcal{A} \subset Z$ is a component of W and the set $\mathcal{N}(\mathcal{A})$ of congestion points of $\mathcal{A}$ is totally disconnected, then $\mathcal{A}$ is path-connected.*

Remark When $W = Z$, $\mathcal{A}$ can be any component of Z.   □

Proof Since $W \subset Z$ is open in Z which is locally compact, W is locally compact and so $\mathcal{A}$ is locally compact because it is closed in W. Since $\mathcal{A}$ is a generalised continuum, if $\mathcal{N}(\mathcal{A}) \neq \emptyset$ all the components of $\mathcal{N}(\mathcal{A})$ are non-degenerate by

Theorem 5.8. Since by hypothesis $\mathcal{N}(\mathscr{A})$ is totally disconnected it follows that $\mathcal{N}(\mathscr{A}) = \emptyset$, and hence $\mathscr{A}$ is path-connected by Lemma 4.23 and Theorem 4.12.

$\square$

The following result is a special case of the lemma.

Theorem 5.26 *Suppose E and F are Banach spaces, and $G : E \to F$ is a function such that its zero set $Z = \{x \in E : G(x) = 0\}$ is locally compact. Let $\mathscr{A} \subset Z$ be a component of a non-empty $W \subset Z$ which is open in Z with the property that the set $\mathcal{N}(\mathscr{A})$ of congestion points of $\mathscr{A}$ is totally disconnected. Then $\mathcal{N}(\mathscr{A}) = \emptyset$ and $\mathscr{A}$ is path-connected.*

The hypotheses of Theorem 5.23 are described in terms of the behaviour of g, whereas it is the properties of the zero set of G that feature in Theorem 5.26. However, for infinite-dimensional nonlinear eigenvalue problems an analogue of Theorem 5.23 follows from the Lyapunov-Schmidt method [9, p. 33], which reduces the original problem to an equivalent problem in finite dimensions. To implement this strategy, additionally terminology is needed.

Definition 5.27 A bounded linear operator $A : E \to F$ is Fredholm operator [5, p. 168] if the dimension $\kappa(A)$ of its kernel is finite and its range is closed in F with finite co-dimension $\rho(A)$. The set $\mathscr{F}(E, F)$ of Fredholm operators is open in the space $\mathscr{L}(E, F)$ of bounded linear operators from E to F and the Fredholm index of A, $\iota(A) = \kappa(A) - \rho(A)$, is continuous in $\mathscr{F}(E, F)$, and hence ι is constant on components of $\mathscr{F}(E, F)$. $\square$

Theorem 5.28 (Lyapunov-Schmidt Reduction)   *Suppose X and Y are Banach spaces and $G : U \to Y$, where U is open in $\mathbb{R} \times X$, satisfies the following:*

(i) $G(\lambda_0, x_0) = 0 \in Y$, $(\lambda_0, x_0) \in U$;
(ii) G is real-analytic on U;
(iii) for all $(\lambda, x) \in U$ the partial derivative $\partial_x G[(\lambda, x)] \in \mathscr{L}(X, Y)$ is a Fredholm operator with index zero.

(a) Let $L = \partial_x G[(\lambda_0, x_0)]$ and $q = \rho(L) = \kappa(L) \neq 0$. Then by the real-analytic version of [7, Thm. 8.2.1], see [7, Rmks. 8.2.2], there exist two open sets $U_0 \subset U$ and $V \subset \mathbb{R} \times \ker(L)$, and two real-analytic mappings $\psi : V \to X$ and $h : V \to \mathbb{R}^q$, such that

$$(\lambda_0, x_0) \in U_0, \ (\lambda_0, 0) \in V, \ \psi(\lambda_0, 0) = x_0, \tag{5.2}$$

and $G(\lambda, x) = 0 \in Y$, $(\lambda, x) \in U_0$ if and only if

$$x = \psi(\lambda, \xi) \text{ where } h(\lambda, \xi) = 0 \text{ for some } (\lambda, \xi) \in V. \tag{5.3}$$

(b) If $L = \partial_x G[(\lambda_0, x_0)]$ and $\rho(L) = \kappa(L) = 0$, by [7, Thm. 4.5.4] there exist an open set $U_0 \subset U$, an open interval $I \subset \mathbb{R}$, $(\lambda_0, x_0) \in U_0$, $\lambda_0 \in I$, and a

real-analytic function $\phi : I \rightarrow X$ such that $G(\lambda, x) = 0$ with $(\lambda, x) \in U_0$ if and only if $\phi(\lambda) = x$ with $\lambda \in I$.   □

Since the Lyapunov-Schmidt method reduces problem (5.2) to an equivalent finite-dimensional problem, namely (5.3) in case (a), and simply $\lambda \in I$ in case (b), the following analogue of Lemma 5.21 is a corollary.

Theorem 5.29 ([15, §1.2, Thm. 1 (i), p. 753]) *If G satisfies (i), (ii) and (iii), the set of solutions of problem (5.2) is locally path-connected at (λ_0, x_0), and hence (λ_0, x_0) is not a congestion point of the solution set.*

To apply Theorem 5.29 to Eq. (1.1) the following classical result from the Fredholm Alternative [5, p. 160] are needed.

When X is a Banach space let $L : X \rightarrow X$ be a compact linear operator and $I : X \rightarrow X$ the identity operator. Then

$$I - L : X \rightarrow X \text{ is a Fredholm operator of index 0 (Definition 5.27)}$$

and in particular

$$\ker(I - \lambda L) = \{0\} \text{ if and only if } \text{range}(I - \lambda L) = X.$$

This is the Fredholm Alternative [5, p. 160].

Path-Connected Global Bifurcation

Consider again the nonlinear eigenvalue problem in Sect. 1.1,

$$x = \lambda L x + R(\lambda, x), \quad (\lambda, x) \in \mathbb{R} \times X, \tag{5.4a}$$

where $L : X \rightarrow X$ is linear and compact, $R : \mathbb{R} \times X \rightarrow X$ is continuous and compact, with $R(\lambda, 0) = 0$ for all $\lambda \in \mathbb{R}$ and

$$\frac{\|R(\lambda, x)\|}{\|x\|} \rightarrow 0 \text{ as } 0 \neq \|x\| \rightarrow 0 \text{ uniformly for } \lambda \text{ in bounded sets.} \tag{5.4b}$$

As in Sect. 1.3 let $\mathcal{T} = \{(\lambda, 0) : \lambda \in \mathbb{R}\}$ denote the line of trivial solutions, S the set of non-trivial solution of (5.4a) and $\mathcal{S} = \overline{S}$.

Remark 5.30 Since $R : \mathbb{R} \times X \rightarrow X$ is continuous and compact, its partial Fréchet derivative $\partial_x R[(\lambda, x)] \in \mathcal{L}(X, X)$, at any point (λ, x) where it exists, is a compact linear operator [28, Lemma 4.1, p. 135].   □

Krasnosel'skii [28, Lemma 2.1, p. 192] observed that λ_0 is a bifurcation point only if λ_0 is a characteristic value of L and proved [28, Thm. 2.1, p. 196] that bifurcation occurs if λ_0 has odd multiplicity. Rabinowitz [40, Thm. 1.3] then used Leray-Schauder degree theory to prove the following global result.

Theorem 5.31 *In* (5.4) *let the above hypotheses on L and R be satisfied and let λ_0 have odd multiplicity. Then there is a set $\mathscr{C}$ in S such that $\overline{\mathscr{C}}$ is a component of $\overline{S} = \mathscr{S}$ with $(\lambda_0, 0) \in \overline{\mathscr{C}}$ and either $\overline{\mathscr{C}}$ is unbounded or there exists $(\lambda_1, 0) \in \overline{\mathscr{C}}$, $\lambda_1 \neq \lambda_0$, and λ_1 is a characteristic value of L of odd multiplicity.*

Now for convenience write equation (5.4a) as $G(\lambda, x) = 0$ where

$$G(\lambda, x) := x - \lambda L x - R(\lambda, x), \quad (\lambda, x) \in \mathbb{R} \times X, \tag{5.5}$$

and note that, at any (λ, x) where it exists, the partial Fréchet derivative of $G : \mathbb{R} \times X \to X$ is given by

$$\partial_x G[(\lambda, x)] := I - \lambda L - \partial_x R[(\lambda, x)] \in \mathscr{L}(X, X),$$

and hence $\partial_x G[(\lambda, x)]$ is a Fredholm operator of index zero by Remark 5.30.

The Main Result

Theorem 5.32 *In the notation of Theorem 5.31 suppose $\mathscr{A}$ is a component of S such that*

$$N_{\mathscr{A}} := \left\{ (\lambda, x) \in \mathscr{A} : R \text{ is not real-analytic at } (\lambda, x) \right\} \text{ is totally disconnected.}$$

Then $\mathscr{A}$ is path-connected and the structure of $\mathscr{A}$ in a neighbourhood of $(\lambda, x) \in \mathscr{A} \setminus N_{\mathscr{A}}$ is given by Theorem 5.28.

Proof With $E = \mathbb{R} \times X$ and $F = X$ in Theorem 5.26, let $Z = \{(\lambda, x) \in \mathbb{R} \times X : G(\lambda, x) = 0\}$ be the zero set of the function G defined by (5.5) and let $W = \mathcal{S}$, the set of non-trivial solutions of $G(\lambda, x) = 0$ which is open in Z. Then it suffices to check that the hypotheses of Theorem 5.26 hold when $\mathscr{A}$ is a component of $\mathcal{S}$ for which $N_{\mathscr{A}}$ is totally disconnected.

First note that since L and R are compact and $\lambda \in \mathbb{R}$, all closed bounded subsets of Z are compact, and hence Z is locally compact in $\mathbb{R} \times X$. Next recall that $\mathscr{N}(\mathcal{S}) \subset \mathscr{N}(Z)$ since $\mathcal{S}$ is open in Z, and hence $\mathscr{N}(\mathcal{S}) \cap \mathscr{A} \subset N_{\mathscr{A}}$ by Theorem 5.29. Moreover, by Lemma 5.18, $\mathscr{N}(\mathscr{A}) \subset \mathscr{N}(\mathcal{S})$, which implies that $\mathscr{N}(\mathscr{A}) \subset N_{\mathscr{A}}$ which, by hypothesis, is totally disconnected. Therefore $\mathscr{N}(\mathscr{A})$ is totally disconnected and, by Theorem 5.26 with $W = \mathcal{S}$, $\mathscr{A}$ is path-connected.

Now if $(\lambda, x) \in \mathscr{A} \setminus N_{\mathscr{A}}$ there is a open neighbourhood U of (λ, x) in $\mathbb{R} \times X$ such that G is real-analytic on U and, by Lemma 5.18, $U \cap \mathcal{S} \subset \mathscr{A}$ (i.e. (λ, x) belongs to the interior of $\mathscr{A}$ in the metric subspace $\mathcal{S}$). Now, since (λ, x) is a solution of (5.2) and $\partial_x R[(\lambda, x)] : X \to X$ is compact (see Remark 5.30), all the hypotheses of Theorem 5.28 are satisfied and hence there is a open neighbourhood $U_0 \subset U$ of

(λ, x) such that $\mathcal{S} \cap U_0$ is given by Theorem 5.28. Then since $\mathcal{A} \cap U_0 = \mathcal{S} \cap U_0$ the result follows.   □

Remark If $(\lambda, x) \in \mathscr{C}$ and $\mathcal{A}$ is the component of $\mathcal{S}$ that contains (λ, x), then $\overline{\mathcal{A}}$ is a connected subset of $\overline{\mathcal{S}}$ that contains (λ, x). Moreover $(\lambda, x) \in \overline{\mathscr{C}}$ which is a component of $\overline{\mathcal{S}}$ by Theorem 5.31. Hence $\overline{\mathcal{A}} \subset \overline{\mathscr{C}}$ and so $\mathcal{A} \subset \overline{\mathscr{C}} \cap \mathcal{S} = \overline{\mathscr{C}} \backslash \mathcal{T} = \mathscr{C}$, by the definition of $\mathscr{C}$ in Theorem 1.3.   □

Chapter 6
Decomposable and Indecomposable Continua

6.1 Decomposable and Indecomposable Continua

Definition 6.1 A non-degenerate continuum is decomposable if it is the union of two of its proper sub-continua (Definition 3.21), and indecomposable otherwise. A degenerate continuum will be considered indecomposable. □

The existence of non-degenerate indecomposable continua in the plane will be proved from first principles in Sect. 6.3.

Remark 6.2 A simple observation, see [23, Thm. 3-51] or [37, pp. 8,9], that a non-degenerate continuum M is indecomposable if it contains three distinct points no pair of which is in a proper sub-continuum of M, is used below in the construction of an indecomposable continuum. However the converse and much more is true, see Corollary 6.6. □

Theorem 6.3 *A non-degenerate continuum M is decomposable if and only it has a proper sub-continuum with non-empty interior.*

Proof Suppose M is decomposable. Then $M = H \cup K$ where H and K are proper sub-continua, and hence $M \setminus H$ is a non-empty open subset of the proper sub-continuum K.

For the contrary, suppose M has a proper sub-continuum K which has non-empty interior. Then $\emptyset \neq \overline{M \setminus K} = M \setminus K^\circ \neq M$ and $M = K \cup \overline{(M \setminus K)}$. Hence, if $M \setminus K$ is connected, $H := \overline{M \setminus K}$ is a proper sub-continuum of M, and M is decomposable since $M = H \cup K$.

If $M \setminus K$ is not connected, since it is open $M \setminus K = U_1 \cup U_2$ where U_1 and U_2 are non-empty disjoint open subsets of M and, since K, U_1 and U_2 are mutually disjoint, it follows that $K \cup U_i$ is closed, $i = 1, 2$. Define

$$H_1 := K \cup U_1, \quad H_2 := K \cup U_2,$$

© The Author(s), under exclusive license to Springer Nature Switzerland AG 2025
B. Buffoni, J. Toland, *Connected Sets in Global Bifurcation Theory*, SpringerBriefs in Mathematics, https://doi.org/10.1007/978-3-031-87051-4_6

so that

$$M = H_1 \cup H_2 = (K \cup U_1) \cup (K \cup U_2),$$

and since $K \cup U_i$ is closed,

$$\overline{U_i} \subset \overline{K \cup U_i} = K \cup U_i \text{ implies } \partial U_i \subset K \text{ for } i = 1, 2.$$

To see that H_i, $i = 1, 2$, is connected, for $x \in U_i$ let C_x^i be the component of $\overline{U_i}$ that contains x. Then from Corollary 3.55 (i) with $A = M$ and $G = U_i$, it follows that $C_x^i \cap \partial U_i \neq \emptyset$ and thus $C_x^i \cup K$ is connected because $C_x^i \cap K \neq \emptyset$ and K is a continuum. Since this holds for every $x \in U_i$,

$$\overline{U_i} \cup K = U_i \cup K \subset \cup_{x \in U_i} (C_x^i \cup K) \subset \overline{U_i} \cup K,$$

and since $\cup_{x \in U_i} (C_x^i \cup K)$ is connected it follows that H_i, $i = 1, 2$ are sub-continua of M. Hence M is decomposable since $M = H_1 \cup H_2$ and $H_1 \neq M \neq H_2$.   □

Remark 6.4 Since proper sub-continua are closed, a non-degenerate continuum M is indecomposable if and only if all its proper sub-continua are nowhere dense in M (Definition 3.3). In particular, a sub-continuum A of M is indecomposable if and only if the interior in A of all proper sub-continua F of A is empty. By (3.7b) this means that a sub-continuum A of M is indecomposable if and only if

$$A \cap (F \cup (M \setminus A))° = \emptyset, \text{ which is stronger than } F° = \emptyset,$$

for all proper closed connected subsets F of A.   □

By Theorem 3.57 (iii), a decomposable continuum $K = K' \cup K''$, where K' and K'' are proper sub-continua of K, coincides with the composant K_p for all $p \in K' \cap K''$. The next result reflects the complexity of a non-degenerate indecomposable continuum.

For p and $q \in K$, write $p \sim q$ when there is a proper sub-continuum of K that contains both p and q. As $\{p\}$ is a proper sub-continuum of K, $p \sim p$, and clearly $p \sim q$ implies $q \sim p$. Thus the relation $\sim$ is reflexive and symmetric.

Theorem 6.5 *When a metric space K is a non-degenerate indecomposable continuum, $\sim$ is an equivalence relation with equivalence classes*

$$\{K_p : p \in K\}, \text{ where } K_p \text{ is the composant of } K \text{ containing } p.$$

Moreover K has uncountably many, mutually disjoint, composants and every point of K is a congestion point.

Remark Note that if K is a non-degenerate indecomposable continuum in a metric space M this means that every point of K is a congestion point of K, not of M.   □

Proof To show that $\sim$ is transitive suppose that $r \sim p$ and $p \sim q$, $r, p, q \in K$, and let K_{rp} and K_{pq} be proper sub-continua of K that contain $\{r, p\}$ and $\{p, q\}$. Then since $q \in K_{rp}$ or $r \in K_{pq}$ would imply $r \sim q$, it suffices to assume that $q \notin K_{rp}$ and $r \notin K_{pq}$. Then since $p \in K_{rp} \cap K_{pq}$, the set $K_{rp} \cup K_{pq}$ is a continuum that is the union of two of its proper sub-continua. Since K is indecomposable, $K_{rp} \cup K_{pq}$ is a proper sub-continuum of K that contains r and q. Therefore $r \sim q$ and so $\sim$ is an equivalence relation. Since $p \sim q$ if and only $q \in K_p$ the equivalence class of p is the composant K_p. Therefore, $K = \cup_{p \in K} K_p$ and the K_ps are mutually disjoint since the equivalence classes of K partition K.

Suppose K is a non-degenerate indecomposable continuum which has only countably many composants. Then, by Theorem 3.57 (ii) each composant is the union of countably many proper sub-continua of K, each sub-continuum having empty interior (Theorem 6.3). It follows from Remark 2.5 that K is the union of countably many nowhere dense sets. But this is false by the Baire Category Theorem 3.12, since K is compact. This shows that K has uncountably many composants.

Finally to prove that every point of K is a congestion point suppose the contrary, that the metric space K is weakly locally connected at $x \in K$. Choose $r > 0$ such that $\overline{B_r(x)} \neq K$ and let V be a connected neighbourhood of x with $V \subset B_r(x)$. Now if W denotes the component of $\overline{B_r(x)}$ that contains x, it follows that W is closed and $x \in V \subset W \neq K$, and since V contains a ball centred at x, W is a proper sub-continuum of K with non-empty interior. This contradicts Theorem 6.3 and the proof is complete. □

Corollary 6.6 *A non-degenerate continuum K is indecomposable if and only if it contains an uncountable set of points no two of which lie in a proper sub-continuum of K.*

Proof If K is a non-degenerate, indecomposable continuum, by Theorem 6.5 there exists an uncountable set $P \subset K$ of points such that $\{K_p : p \in P\}$ consists of disjoint composants. Consequently no pair from P can belong to the same proper sub-continuum of K. Conversely, if K contains one triple $\{r, p, q\}$ no pair of which belongs to the same proper sub-continuum of K, it is clear that K is indecomposable. □

Corollary 6.7 *For a non-degenerate, indecomposable continuum K, let $G \subset K$ be an open set in K with $\emptyset \neq \overline{G} \neq K$. Then $\overline{G}$ has uncountably many non-degenerate components each of which intersect G and ∂G.*

Proof Let $P \subset K$ be such that $\{K_p : p \in P\}$ is the uncountable set of mutually disjoint composants of K given by Theorem 6.5. Then by Theorem 3.57, for all $p \in P$ there exists $x_p \in G \cap K_p$, and the component C_p of $\overline{G}$ which contains x_p, is closed since $\overline{G}$ is closed. It follows that C_p is a proper sub-continuum of K. Hence $C_p \subset K_p$ since $x_p \in K_p \cap C_p$ and C_p is non-degenerate because $x_p \in C_p \cap G$ and $C_p \cap \partial G \neq \emptyset$ by Corollary 3.55 (i). The sets C_p, $p \in P$, are mutually disjoint because $C_p \subset K_p$, and the sets K_p, $p \in P$, are disjoint. This completes the proof. □

Lemma 6.8 *A non-degenerate path in any metric space M is decomposable.*

Proof Let $K = \{f(t) : t \in [0, 1]\}$ be a non-degenerate path in M. Then, with the metric inherited from M, K is a continuum which has no congestion points since K is locally connected by Lemma 4.10. Therefore K is decomposable since by Theorem 6.5 every point of a non-degenerate indecomposable continuum is a congestion point.   □

Definition 6.9 If a non-degenerate continuum $\mathcal{P}$ and all its proper sub-continua are indecomposable, $\mathcal{P}$ is hereditarily indecomposable.   □

Lemma 6.10 *A hereditarily indecomposable continuum $\mathcal{P}$ contains no non-trivial paths.*

Proof Since by Lemma 6.8 a non-trivial path in $\mathcal{P}$ is a decomposable subcontinuum, $\mathcal{P}$ contains no non-trivial paths if $\mathcal{P}$ is hereditarily indecomposable.   □

Theorem 6.11 (Knaster [25], see also [23, p. 142-3]) *There is a non-degenerate, hereditarily indecomposable continuum $\mathcal{P}$ in $\mathbb{R}^2$.*

After Knaster's example Mazurkiewicz [33] showed that non-degenerate, hereditarily indecomposable continua are not rare by proving that they are of second Baire category in a complete Hausdorff metric space (see Lemma 3.47 in Sect. 3.9).

Then Bing [3] and Moise [34] constructed a non-degenerate, hereditarily indecomposable continuum as the intersection of an infinite family of nested continua defined by simple chains joining points in $\mathbb{R}^2$, and the result was called a pseudo-arc by Moise. The approach of [3, 34] led to a theory of hereditarily indecomposable continua in more general settings and a vast literature, see [31].

For a proof of the existence of hereditarily indecomposable continua in the plane that does not rely explicitly on nested "crooked" sub-chains as in [37, p. 13], see Sect. 6.4, and for a related proof of Mazurkiewicz's result about Baire Categories, see Sect. 6.5.

6.2 Disc Chains

Definition 6.12 A Jordan curve J is the image of an injective continuous map of a circle into $\mathbb{R}^2$.   □

Theorem 6.13 (The Jordan Curve Theorem) *The complement $\mathbb{R}^2 \setminus J$ of a Jordan curve has exactly two components, one bounded and one unbounded. Moreover J is the boundary of both components.*

For a proof of this famous result, see [38, Thm. 10.2] or [51, p. 63]. In the present chapter it will be applied to piecewise-smooth Jordan curves for which the two components in the statement can be described explicitly, and the proof would be correspondingly simpler.

Definition 6.14 In $\mathbb{R}^2$ with the standard Euclidean metric, the ball $B_r(x)$ will be referred to as the open disc with centre x and radius r, denoted $D_r(x)$, and $\mathbf{D}_\epsilon$ denotes the set of discs with radius strictly less than ϵ. □

Definition 6.15 In an open set $A \subset \mathbb{R}^2$, a disc chain $\mathscr{D} = \{D_1, \cdots, D_n\}$ is a simple chain in A (Definition 3.29) with links $D_i = D_{\epsilon_i}(x_i) \subset A$.

If in addition, with respect to closures in $\mathbb{R}^2$,

$$n \geqslant 2, \quad D_1 \not\subset D_2, \quad D_n \not\subset D_{n-1}, \quad \overline{D_i} \cap \overline{D_j} = \emptyset \text{ for all } |i - j| \geqslant 2,$$

then $\mathscr{D}$ will be called a strict disc chain in A.

For distinct points $x, y, z_1, \cdots, z_n \in A$, a disc chain is said to join x to y via $\{z_1, \cdots, z_n\}$ if (see (3.10)) $x \in D_1$ only, $y \in D_n$ only, and $z_i \in [\mathscr{D}]$ for all i, where $[\mathscr{D}]$ is defined by (3.9b). □

Now, for a disc chain $\mathscr{D} = \{D_{\epsilon_1}(x_1), \cdots, D_{\epsilon_n}(x_n)\}$ and $\sigma \in (0, 1)$, let $\mathscr{D}_\sigma$ denote the set $\{D_{\sigma\epsilon_1}(x_1), \cdots, D_{\sigma\epsilon_n}(x_n)\}$. Note that if $\mathscr{D} \subset \mathbf{D}_\epsilon$, then $\mathscr{D}_\sigma \subset \mathbf{D}_{\sigma\epsilon}$, but when $\sigma > 0$ is small, $\mathscr{D}_\sigma$ is not a disc chain. However, when $\sigma < 1$ is sufficiently close to 1 it is a disc chain.

Moreover, if $D_1 \not\subset D_2$, $D_n \not\subset D_{n-1}$, $n \geqslant 2$ and $\sigma < 1$ is sufficiently close to 1, the disc chain $\mathscr{D}_\sigma$ is strict. Furthermore, if $\mathscr{D}$ is already a strict disc chain and the definition of $\mathscr{D}_\sigma$ is extended to values of $\sigma > 1$, $\mathscr{D}_\sigma$ is a strict disc chain for all $\sigma > 1$ sufficiently close to 1.

Definition 6.16 $\mathscr{D}_\sigma$ will be referred to as a reduction of $\mathscr{D}$ if $\sigma < 1$ is sufficiently close to 1 that $\mathscr{D}_\sigma$ is a disc chain, and the notation $\mathscr{D}_\sigma$ implies that meaning. □

Remark 6.17 Suppose $D_\epsilon(x)$ is a disc in A where A is open and connected. Then for $\sigma \in (0, 1)$, its boundary $\partial D_{\sigma\epsilon}(x)$ is a circle in A which is a Jordan curve. Moreover, if $\mathscr{D}$ is a strict disc chain in A, the boundary $\partial[\mathscr{D}_\sigma]$ of $[\mathscr{D}_\sigma]$ with $\sigma \in (0, 1)$ close enough to 1 is a Jordan curve in A which is the union of a finite set of closed arcs of circles, and the bounded component of its complement in $\mathbb{R}^2$ is $[\mathscr{D}_\sigma]$. □

Lemma 6.18 *Suppose $A \subset \mathbb{R}^2$ is open, connected and $D_\epsilon(x) \subset A$. Then for $\sigma \in (0, 1)$, $A \setminus \overline{D_{\sigma\epsilon}(x)}$ is non-empty, open, path-connected, and hence connected. The result also holds for $\sigma = 1$ if $\overline{D_\epsilon(x)} \subset A$.*

Proof Since $A \setminus \overline{D_{\sigma\epsilon}(x)}$ is open and non-empty when $\sigma \in (0, 1)$, for distinct points $y, z \in A \setminus \overline{D_{\sigma\epsilon}(x)}$, there exists $\nu \in (\sigma\epsilon, \epsilon)$ such that $\nu < \min\{|y - x|, |z - x|\}$. Since A is path-connected by Lemma 4.2, there are two paths in A which join y to x and z to x, respectively, and by continuity of the distance function both intersect the circle $C = \{x' \in \mathbb{R}^2 : |x' - x| = \nu\}$. Hence there are two paths in $A \setminus \overline{D_{\sigma\epsilon}(x)}$ which join both y and z to points in C. Since C is a path-connected subset of $A \setminus \overline{D_{\sigma\epsilon}(x)}$, it follows that y and z are connected by a path in $A \setminus \overline{D_{\sigma\epsilon}(x)}$. This shows that $A \setminus \overline{D_{\sigma\epsilon}(x)}$ is path-connected, and hence connected. If $\overline{D_\epsilon(x)} \subset A$, a similar proof yields the result for $\sigma = 1$. □

Lemma 6.19 *If A is open and connected and $\mathscr{D}_\sigma$ is a reduction of a disc chain $\mathscr{D}$ in A, then $A \setminus \overline{[\mathscr{D}_\sigma(k,m)]}$ is connected for any sub-chain $\mathscr{D}_\sigma(k,m)$.*

Proof It is sufficient to prove the result for $A \setminus \overline{[\mathscr{D}_\sigma]}$, the argument for sub-chains of $\mathscr{D}_\sigma$ being the same. The idea is to adapt the preceding proof to this more general setting. Since $A \setminus \overline{[\mathscr{D}_\sigma]}$ is open and non-empty, for distinct points $y, z \in A \setminus \overline{[\mathscr{D}_\sigma]}$ let $\nu \in (\sigma, 1)$ be such that $y, z \notin \overline{[\mathscr{D}_\nu]}$. Since A is path-connected, by Lemma 4.2 for any point $a \in [\mathscr{D}_\nu]$ there are two paths in A joining y and z to a.

Now, by Remark 6.17, $\partial[\mathscr{D}_\nu]$ is a Jordan curve, $a \in [\mathscr{D}_\nu]$ is in the bounded component of its complement and y and z are in the unbounded component. Since paths are connected, each of the two paths in A joining y and z to a must intersect $\partial[\mathscr{D}_\nu]$ and, since $\partial[\mathscr{D}_\nu]$ is path-connected and $\partial[\mathscr{D}_\nu] \subset A \setminus \overline{[\mathscr{D}_\sigma]}$, it follows that y and z are connected by a path in $A \setminus \overline{[\mathscr{D}_\sigma]}$. Therefore $A \setminus \overline{[\mathscr{D}_\sigma]}$ is path-connected, and hence connected. □

Lemma 6.20 *For $\widetilde{n}, n \geqslant 2$ let $\widetilde{\mathscr{D}} = \{\widetilde{D}_1, \cdots, \widetilde{D}_{\widetilde{n}}\}$ and $\mathscr{D} = \{D_1, \cdots, D_n\}$ be strict disc chains in $\mathbb{R}^2$ such that $D_1 = \widetilde{D}_1$ and $\overline{[\mathscr{D}(2,n)]} \subset \overline{[\widetilde{\mathscr{D}}]}$.*

Then $\overline{[\widetilde{\mathscr{D}}]} \setminus \overline{[\mathscr{D}(1,m)]}$ is connected for $m \in \{2, \ldots, n\}$, and for $m \in \{2, \ldots, n\}$, $\overline{[\widetilde{\mathscr{D}}]} \setminus \left(\overline{[\mathscr{D}(1,m)]} \cup \overline{\widetilde{D}_{\widetilde{n}}}\right)$ is connected if $\overline{[\mathscr{D}(2,n)]} \subset \overline{[\widetilde{\mathscr{D}}]} \setminus \overline{\widetilde{D}_{\widetilde{n}}}$.

Proof It suffices to prove the result for $m = n$, the other cases being essentially the same. Let $\mathscr{D} = \{D_{\epsilon_1}(x_1), \cdots, D_{\epsilon_n}(x_n)\}$ and, for $\tau > 1$,

$$\mathscr{D}^\tau = \{D_{\epsilon_1}(x_1), D_{\tau\epsilon_2}(x_2), \cdots, D_{\tau\epsilon_n}(x_n)\}.$$

Since $\mathscr{D}$ is a strict disk chain, for $\tau > 1$ close to 1, $\mathscr{D}^\tau$ is also a strict disc chain and

$$\overline{D_{\tau\epsilon_2}(x_2)} \subset \overline{[\widetilde{\mathscr{D}}]}, \ldots, \overline{D_{\tau\epsilon_n}(x_n)} \subset \overline{[\widetilde{\mathscr{D}}]}.$$

As in Remark 6.17, $\partial[\mathscr{D}^\tau]$ is a Jordan curve and $B := \overline{[\widetilde{\mathscr{D}}]} \cap \partial[\mathscr{D}^\tau]$ is a non-empty, path-connected, set obtained from $\partial[\mathscr{D}^\tau]$ by removing an arc of the circle $\partial\widetilde{D}_1$.

Let $y, z \in \overline{[\widetilde{\mathscr{D}}]} \setminus \overline{[\mathscr{D}]}$ and choose $\tau > 1$ near 1 such that $y, z \notin \overline{[\mathscr{D}^\tau]}$. Since $\overline{[\widetilde{\mathscr{D}}]}$ is path-connected by Lemma 4.2, there are two paths in $\overline{[\widetilde{\mathscr{D}}]}$ which join both y and z to a single point in $[\mathscr{D}] \subset [\mathscr{D}^\tau]$ and, since both paths intersect the boundary $\partial[\mathscr{D}^\tau]$, they intersect $B = \overline{[\widetilde{\mathscr{D}}]} \cap \partial[\mathscr{D}^\tau]$, paths being connected. Since B is a path-connected subset of $\overline{[\widetilde{\mathscr{D}}]} \setminus \overline{[\mathscr{D}]}$, it follows that y and z are connected by a path in $\overline{[\widetilde{\mathscr{D}}]} \setminus \overline{[\mathscr{D}]}$. This shows that $\overline{[\widetilde{\mathscr{D}}]} \setminus \overline{[\mathscr{D}]}$ is path-connected, and hence connected.

Finally, since $\overline{[\widetilde{\mathscr{D}}]} \setminus \overline{\widetilde{D}_{\widetilde{n}}}$ is path-connected, if $\overline{[\mathscr{D}(2,n)]} \subset \overline{[\widetilde{\mathscr{D}}]} \setminus \overline{\widetilde{D}_{\widetilde{n}}}$ the same argument shows $\overline{[\widetilde{\mathscr{D}}]} \setminus \left(\overline{[\mathscr{D}(1,m)]} \cup \overline{\widetilde{D}_{\widetilde{n}}}\right)$ is connected, $m \in \{2, \ldots, n\}$. □

The next result is the analogy for strict disc chains of Corollary 3.34, which implies that when $A \subset \mathbb{R}^2$ is open and connected, $x, y \in A$ and $\epsilon > 0$, there is a disc chain in A joining x to y with links in $\boldsymbol{D}_\epsilon$.

Lemma 6.21 *Suppose $A \subset \mathbb{R}^2$ is open and connected, $\epsilon > 0$, and c_E and c_F are the centres of discs E and F with*

$$\overline{E} \cap \overline{F} = \emptyset, \quad E \cap A \neq \emptyset \text{ and } F \cap A \neq \emptyset.$$

Then there is a strict disc chain $\mathscr{D} = \{D_1, \cdots, D_n\}$ in $\mathbb{R}^2$ which joins c_E to c_F with

$$n \geqslant 3, \quad D_1 = E, \quad D_n = F, \quad c_E \notin \overline{D_2}, \quad c_F \notin \overline{D_{n-1}}, \quad \overline{\cup_{i=2}^{n-1} D_i} \subset A$$

and, for all $i \in \{2, \ldots, n-1\}$, $D_i \in \boldsymbol{D}_\epsilon$. If in addition $\overline{E} \cup \overline{F} \subset A$, $\overline{[\mathscr{D}]} \subset A$.

Proof Let $\widetilde{A} = A \cup E \cup F$, which is open and connected, be covered in $(\widetilde{A}, d_{\widetilde{A}})$ by the following family of open discs:

$$\mathscr{G} = \{E, F\} \bigcup \left\{ D_\delta(z) : \delta \in (0, \epsilon), \overline{D_\delta(z)} \subset A, \{c_E, c_F\} \cap \overline{D_\delta(z)} = \emptyset, z \in A \right\}.$$

By Theorem 3.33, there exists $n \geqslant 3$, $\{x_1, \cdots, x_n\} \subset \widetilde{A}$ and $\{\epsilon_1, \cdots, \epsilon_n\} \subset (0, \infty)$ with $D_{\epsilon_1}(x_1) = E$, $D_{\epsilon_n}(x_n) = F$, $c_E \in \overline{D_{\epsilon_i}(x_i)}$ only if $i = 1$, $c_F \in \overline{D_{\epsilon_i}(x_i)}$ only if $i = n$ $(1 \leqslant i \leqslant n)$, $\overline{D_{\epsilon_i}(x_i)} \subset A$, $\epsilon_i < \epsilon$ for all $1 < i < n$, and

$$D_{\epsilon_i}(x_i) \cap D_{\epsilon_j}(x_j) \neq \emptyset \text{ if and only if } |i - j| \leq 1, \quad 1 \leq i, j \leq n.$$

Finally, for each $1 < i < n$ replace, if necessary, the value of ϵ_i by a slightly smaller one so that the simple chain becomes strict and then, for each $1 \leqslant i \leqslant n$, set $D_i = D_{\epsilon_i}(x_i)$. $\qquad\square$

Theorem 6.22 *For $\epsilon > 0$ and distinct points $\alpha, \beta, \gamma \in A$, where A is open and connected in $\mathbb{R}^2$, there is a disc chain $\mathscr{D}$ with links in $\boldsymbol{D}_\epsilon$ and $\overline{[\mathscr{D}]} \subset A$, which joins α to β via γ, see Definition 6.15.*

Proof For $0 < r < \frac{1}{4} \min\{|\alpha - \beta|, |\alpha - \gamma|, |\beta - \gamma|\}$, let $D_r(\alpha)$, $D_r(\beta)$ and $D_r(\gamma)$ be open discs such that $\overline{D_r(\alpha)} \cup \overline{D_r(\beta)} \cup \overline{D_r(\gamma)} \subset A$ and, by Lemma 6.18, $A \setminus \overline{D_r(\beta)}$ is connected. Then for $\epsilon \in (0, r)$, by Lemma 6.21 there is a disc chain $\widetilde{\mathscr{D}} = \{\widetilde{D}_1, \cdots, \widetilde{D}_n\}$, with $n \geqslant 3$ links in $\boldsymbol{D}_\epsilon$ which joins α to γ in $A \setminus \overline{D_r(\beta)}$, with $\widetilde{D}_1 = D_{\epsilon/2}(\alpha)$ and $\widetilde{D}_n = D_{\epsilon/2}(\gamma)$. Moreover, by Lemma 6.19, $\widetilde{\mathscr{D}}$ can be chosen such that $A \setminus [\widetilde{\mathscr{D}}(1, n-1)]$ is open and connected, and $\overline{[\widetilde{\mathscr{D}}]} \subset A$.

Now since $\widetilde{D}_n \subset \widehat{A}$ and $D_{\epsilon/2}(\beta) \subset \widehat{A}$ where $\widehat{A} := \widetilde{D}_n \cup (A \setminus [\widetilde{\mathscr{D}}(1, n-1)])$ is open and connected, by Lemma 6.21, there is a disc chain $\widehat{\mathscr{D}} = \{\widehat{D}_1, \cdots, \widehat{D}_m\}$ in $\widehat{A}$ joining γ to β, with $m \geqslant 3$ links in $\boldsymbol{D}_\epsilon$ and $\widehat{D}_1 = \widetilde{D}_n$ and $\widehat{D}_m = D_{\epsilon/2}(\beta)$. Therefore $\mathscr{D} = \{\widetilde{D}_1, \ldots, \widetilde{D}_{n-1}, \widetilde{D}_n = \widehat{D}_1, \widehat{D}_2, \ldots, \widehat{D}_m\}$ is a disc chain with $n + m - 1$ links joining α to β via γ, $\overline{[\mathscr{D}]} \subset A$ and the only link containing γ is $\widetilde{D}_n = \widehat{D}_1$. $\qquad\square$

6.3   An Indecomposable Continuum in the Plane

Theorem 6.23 *There exists a non-degenerate, indecomposable continuum in the plane.*

Proof For distinct points $\alpha, \beta, \gamma \in \mathbb{R}^2$, by Theorem 6.22 there exists a disc chain $\mathcal{G}_1$ joining α to γ via β, with links of radius less than 1. Now let $\mathcal{G}_2$ be a disc chain with $\overline{[\mathcal{G}_2]} \subset [\mathcal{G}_1]$ joining β to γ via α with links of radius less than $1/2$, and let $\mathcal{G}_3$ be a disc chain with $\overline{[\mathcal{G}_3]} \subset [\mathcal{G}_2]$ joining α to β via γ, with links of radius less than $1/2^2$.

To appreciate the complexity of this construction see [23, Fig. 3-21, p. 141] or [37, Fig. 1.10, p. 8], or just sketch $\mathcal{G}_2$ and $\mathcal{G}_3$ when $\mathcal{G}_1 = \{G_1, \cdots, G_5\}$ is a disc chain with five discs
$$G_i = B_i\big((i, 0), 3/4\big), 1 \leqslant i \leqslant 5, \text{ where } \alpha = (1, 0), \beta = (3, 0), \gamma = (5, 0).$$

By Theorem 6.22 and induction there is a sequence $\mathcal{G}_k$, $k \in \mathbb{N}$, of disc chains with links of radius less than 2^{1-k} and $\overline{[\mathcal{G}_{k+1}]} \subset [\mathcal{G}_k]$ so that, for $n \in \mathbb{N} \cup \{0\}$,

$$\mathcal{G}_k \text{ joins } \alpha \text{ to } \gamma \text{ via } \beta \text{ when } k = 3n + 1;$$

$$\mathcal{G}_k \text{ joins } \beta \text{ to } \gamma \text{ via } \alpha \text{ when } k = 3n + 2;$$

$$\mathcal{G}_k \text{ joins } \alpha \text{ to } \beta \text{ via } \gamma \text{ when } k = 3n + 3.$$

Let $K = \cap_{k \in \mathbb{N}}[\mathcal{G}_k] = \cap_{k \in \mathbb{N}}\overline{[\mathcal{G}_k]}$. Then since $[\mathcal{G}_{k+1}] \subset [\mathcal{G}_k]$ it is immediate that

$$\begin{aligned} K &= \cap_{k \in \mathbb{N}}[\mathcal{G}_{3k+1}] = \cap_{k \in \mathbb{N}}[\mathcal{G}_{3k+2}] = \cap_{k \in \mathbb{N}}[\mathcal{G}_{3k+3}] \\ &= \cap_{k \in \mathbb{N}}\overline{[\mathcal{G}_{3k+1}]} = \cap_{k \in \mathbb{N}}\overline{[\mathcal{G}_{3k+2}]} = \cap_{k \in \mathbb{N}}\overline{[\mathcal{G}_{3k+3}]}, \end{aligned} \tag{6.1}$$

from which it follows that $\{\alpha, \beta, \gamma\} \subset K$, and K is a non-degenerate continuum by Theorem 3.43. By Remark 6.2, K will be indecomposable if no proper sub-continuum of K contains any of the sets $\{\alpha, \gamma\}, \{\alpha, \beta\}$ or $\{\beta, \gamma\}$.

So suppose that H is a proper sub-continuum of K which contains $\{\alpha, \gamma\}$ and let $\rho \in K \setminus H$. Then dist $(\rho, H) > 0$ and if $2^{-3N} < \text{dist}(\rho, H)/2$ the links of $\mathcal{G}_{3N+1}$ which contain ρ do not intersect H. Now by construction α is in the first, and only the first, link of $\mathcal{G}_{3N+1}$ and similarly γ is in the last, and only the last, link of $\mathcal{G}_{3N+1}$. Moreover, since $\rho \in K$ it follows that $\mathcal{G}_{3N+1}$ is a disc chain from α to γ via ρ, and ρ does not belong to the first or last link because by assumption $\alpha, \gamma \in H$.

Say $\mathcal{G}_{3N+1} = \{G_1, \cdots, G_p, \cdots G_m\}$ where $\rho \in G_p$. Then because $\mathcal{G}_{3N+1}$ is a disc chain, $H \subset [\mathcal{G}_{3N+1}], G_p \cap H = \emptyset, \alpha \in G_1$ and $\gamma \in G_m$, it follows from Lemma 3.32 that H is not connected, which is false. Therefore there is no proper sub-continuum of K which contains $\{\alpha, \gamma\}$. That there is no proper sub-continuum containing β and γ follows by the same argument with $\mathcal{G}_{3N+2}$ replacing $\mathcal{G}_{3N+1}$, and there is no proper sub-continuum containing α and β follows with $\mathcal{G}_{3N+1}$ replaced by $\mathcal{G}_{3N+3}$. Therefore, K in (6.1) is indecomposable by Remark 6.2.   □

6.4   Hereditarily Indecomposable Continua in the Plane

This section is devoted to a variant of the approach of Bing [3] and Moise [34] to the proof of Theorem 6.25, and hence of Theorem 6.11.

Here nested disc chains will be used, but it will not be required as in [3, 34] that each chain is "crooked" [37, p. 13] in the preceding one. Instead a nested family of chains indexed by the set $\mathfrak{W}$ in (6.2) will be constructed so as to have zig-zag behaviour less demanding than crookedness (see Fig. 6.1). For similar ideas, see [26, 27, 39] and [2, 21, 24].

Lemma 6.24 *Let* $\mathcal{V} = \{V_1, \cdots, V_n\}$, $n \geqslant 4$, *be a strict disc chain joining* $a_1 \in V_1$ *to* $a_n \in V_n$. *Then there is a strict disc chain* $\mathcal{U} = \{U_1, \cdots, U_m\}$ *joining* a_1 *to* a_n *such that*

$$U_1 = V_1, \quad U_m = V_n, \quad \overline{U_i} \subset [\mathcal{V}(2, n-1)] \text{ for all } i \in \{2, \ldots, m-1\},$$

and there exist $1 < i < j < m$ *satisfying*

$$\overline{U_i} \subset V_{n-1} \text{ and } \overline{U_j} \subset V_2.$$

Proof Choose $a_2 \in V_2 \backslash (\overline{V_1} \cup \overline{V_3})$ and $a_{n-1} \in V_{n-1} \backslash (\overline{V_{n-2}} \cup \overline{V_n})$, and let $r > 0$ be small enough, so that

$$\overline{D_r(a_2)} \subset V_2 \backslash (\overline{V_1} \cup \overline{V_3}), \quad \overline{D_r(a_{n-1})} \subset V_{n-1} \backslash (\overline{V_{n-2}} \cup \overline{V_n}).$$

As $[\mathcal{V}(2, n-1)] \backslash \overline{V_n}$ is open and connected, Lemma 6.18 ensures that

$$A_1 := [\mathcal{V}(2, n-1)] \backslash (\overline{D_r(a_2)} \cup \overline{V_n})$$

is open and connected. Since

$$V_1 \cap A_1 \neq \emptyset \text{ and } \overline{D_r(a_{n-1})} \subset A_1,$$

by Lemma 6.21, there is a strict disc chain $\mathcal{D}_1 = \{D_1^1, \cdots, D_{n_1}^1\}$ such that

$$n_1 \geqslant 3, \quad D_1^1 = V_1, \quad D_{n_1}^1 = D_r(a_{n-1}), \quad \overline{[\mathcal{D}_1(2, n_1-1)]} \subset A_1.$$

By the last sentence of Lemma 6.20,

$$A_2 := [\mathcal{V}] \backslash (\overline{[\mathcal{D}_1(1, n_1-1)]} \cup \overline{V_n})$$

is open and connected. Since

$$D_r(a_{n-1}) \cap A_2 \neq \emptyset \text{ and } \overline{D_r(a_2)} \subset A_2,$$

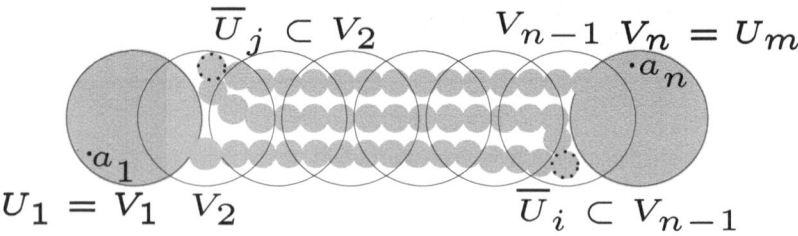

Fig. 6.1 In this schematic illustration of the proof of Lemma 6.24, the black circles represent the links of the strict disc chain $\mathscr{V}$ and the gray discs denote the links of the strict disc chain $\mathscr{U}$. In the proof, the first sub-chain of $\mathscr{U}$ joins V_1 to a gray disc $D_r(a_{n-1}) \subset \overline{D_r(a_{n-1})} \subset V_{n-1}\setminus(\overline{V_{n-2}} \cup \overline{V_n})$ with black dots on its circumference, the second sub-chain of $\mathscr{U}$ joins $D_r(a_{n-1})$ to a gray disc $D_r(a_2) \subset \overline{D_r(a_2)} \subset V_2\setminus(\overline{V_1} \cup \overline{V_3})$ with black dots on its circumference, while the third sub-chain of $\mathscr{U}$ joins $D_r(a_2)$ to V_n. Note that although $D_r(a_{n-1})$ and $D_r(a_2)$ play the role of U_i and U_j in the lemma, U_i and U_j are not uniquely determined by its conclusions. The links of $\mathscr{U}$, or of $\mathscr{V}$, may have different radii

by Lemma 6.21, there is a strict disc chain $\mathscr{D}_2 = \{D_1^2, \cdots, D_{n_2}^2\}$ joining a_{n-1} to a_2 such that

$$n_2 \geqslant 3, \quad D_1^2 = D_r(a_{n-1}), \quad D_{n_2}^2 = D_r(a_2), \quad \overline{[\mathscr{D}_2(2, n_2 - 1)]} \subset A_2 \,.$$

As $\overline{\mathscr{D}_1(1, n_1 - 1)} \cap \overline{\mathscr{D}_2(2, n_2)} = \emptyset$, observe that $\mathscr{D}_1 \cup \mathscr{D}_2$ is a strict disc chain joining a_1 to a_2 via a_{n-1}. Moreover

$$\overline{[\mathscr{D}_1(2, n_1)]} \cup \overline{[\mathscr{D}_2]} \subset A_1 \cup A_2 \subset [\mathscr{V}(2, n - 1)]\setminus\overline{V_n} \,.$$

By Lemma 6.20,

$$A_3 := [\mathscr{V}(1, n - 1)] \setminus \left(\overline{[\mathscr{D}_1]} \cup \overline{[\mathscr{D}_2(1, n_2 - 1)]} \right)$$

is open and connected. Since

$$D_r(a_2) \cap A_3 \neq \emptyset \quad \text{and} \quad V_n \cap A_3 \neq \emptyset,$$

by Lemma 6.21, there is a strict disc chain $\mathscr{D}_3 = \{D_1^3, \cdots, D_{n_3}^3\}$ joining a_2 to a_n such that

$$n_3 \geqslant 3, \quad D_1^3 = D_r(a_2), \quad D_{n_3}^3 = V_n, \quad \overline{[\mathscr{D}_3(2, n_3 - 1)]} \subset A_3 \,.$$

Therefore $\mathscr{U} = \mathscr{D}_1 \cup \mathscr{D}_2 \cup \mathscr{D}_3$ is a strict disc chain with $m := n_1 + n_2 + n_3 - 2$ links joining a_1 to a_n via a_{n-1} and a_2. Moreover

$$\overline{D_{n_1}^1} = \overline{D_1^2} = \overline{D_r(a_{n-1})} \subset V_{n-1} \quad \text{and} \quad \overline{D_{n_2}^2} = \overline{D_1^3} = \overline{D_r(a_2)} \subset V_2.$$

Finally, in the statement of the lemma let $i = n_1$ and $j = n_1 + n_2 - 1$. Then $U_1 = D_1^1 = V_1$, $U_m = D_{n_3}^3 = V_n$ and $\overline{U_i} \subset A_1 \cup A_2 \cup A_3 \subset [\mathcal{V}(2, n-1)]$ for all $i \in \{2, \ldots, m-1\}$. This completes the proof. $\qquad\square$

The Index Set $\mathfrak{W}$ For a countable base $\mathcal{W}$ of $\mathbb{R}^2$, let $\mathfrak{W}$ be an enumeration of all subsets of $\mathcal{W}$ with exactly two non-empty elements with disjoint closures. In other words

$$\mathfrak{W} = \left\{ \{W_{1,k}, W_{2,k}\} \subset \mathcal{W} : W_{1,k} \neq \emptyset \neq W_{2,k}, \ \overline{W_{1,k}} \cap \overline{W_{2,k}} = \emptyset, \ k \in \mathbb{N} \right\}. \tag{6.2}$$

Note that elements of $\mathfrak{W}$ are not ordered pairs and $\epsilon_k \to 0$ as $k \to \infty$ where

$$\epsilon_k = \min\{1/k, \ \mathrm{dist}\,(W_{1,k}, W_{2,k})/6\} > 0, \quad k \in \mathbb{N}. \tag{6.3}$$

For fixed $a \neq b \in \mathbb{R}^2$ there exists a strict disc chain $\mathcal{G}_0$ joining a to b in $\mathbb{R}^2$ with $n_0 \geqslant 2$ links, for example $\mathcal{G}_0 = \{D_r(a), D_r(b)\}$ where $r = \frac{3}{4}\,\mathrm{dist}\,(a, b)$. Then to proceed by induction suppose, for $k \in \mathbb{N}$, that $\mathcal{G}_{k-1}$ in $\mathbb{R}^2$ is a strict disc chain with n_{k-1} links joining a to b. Then another strict disc chain $\mathcal{H}_k$ joining a to b can be defined as follows.

First, with ϵ_k in (6.3), let $\mathcal{H}_k = \{H_1^k, \cdots, H_{n_k'}^k\}$ be a strict disc chain given by Lemma 6.21 with $A = [\mathcal{G}_{k-1}]$, $c_E = a$ and $c_F = b$, that joins a to b, with n_k' links in D_{ϵ_k} such that

$$[\mathcal{H}_k] \subset [\mathcal{G}_{k-1}], \quad H_1^k = D_\rho(a) \text{ and } H_{n_k'}^k = D_\rho(b),$$

where $\rho \in (0, \epsilon_k)$ is so small that

$$\overline{D_\rho(a)} \cup \overline{D_\rho(b)} \subset [\mathcal{G}_{k-1}] \text{ and } \overline{D_\rho(a)} \cap \overline{D_\rho(b)} = \emptyset.$$

Then note from (6.3) that if $p_1, p_2 \in \{1, \ldots, n_k'\}$ and $\{i, j\} \subset \{1, 2\}$, $i \neq j$, are such that

$$p_1 \leqslant p_2, \ H_{p_1}^k \cap W_{i,k} \neq \emptyset \text{ and } H_{p_2}^k \cap W_{j,k} \neq \emptyset, \text{ then } p_2 \geqslant p_1 + 3. \tag{6.4}$$

In particular if $\mathcal{H}_k(p_1, p_2)$ is a disc sub-chain of $\mathcal{H}_k$ with

$$\left. \begin{array}{c} H_{p_1}^k \cap W_{1,k} \neq \emptyset, \ H_{p_2}^k \cap W_{2,k} \neq \emptyset, \\ \text{and } H_p^k \cap (W_{1,k} \cup W_{2,k}) = \emptyset \text{ for all } p \text{ strictly between } p_1 \text{ and } p_2, \end{array} \right\} \tag{6.5}$$

it follows that $|p_1 - p_2| \geqslant 3$.

Then by Lemma 6.24, $\mathcal{H}_k(p_1, p_2)$ can be replaced by a new sub-chain (see Fig. 6.1) that has the same end links, $H_{p_1}^k$ and $H_{p_2}^k$, as $\mathcal{H}_k(p_1, p_2)$ and by (6.5) no links except the end links of the replacement sub-chain intersect the base sets $W_{1,k}$ and $W_{2,k}$.

Now if all the sub-chains satisfying (6.5) are replaced in this way, and all other discs of $\mathcal{H}_k$ are left undisturbed, when the replaced sub-chains and undisturbed discs are arranged in the order in which they occur in $\mathcal{H}_k$ and re-labelled accordingly, the result is a strict disc chain $\mathcal{G}_k = \{G_1^k, \cdots, G_{n_k}^k\}$ joining a to b with the following properties:

(A) $[\mathcal{G}_k] \subset [\mathcal{H}_k] \subset \overline{[\mathcal{H}_k]} \subset [\mathcal{G}_{k-1}]$, $k \in \mathbb{N}$, and $\mathcal{G}_k = \{G_1^k, \ldots, G_{n_k}^k\}$ joins a to b with n_k links;

(B) if $(x, y) \in (W_{1,k} \times W_{2,k}) \cup (W_{2,k} \times W_{1,k})$ and $1 \leqslant p_1 < p_2 \leqslant n_k'$ satisfy

$$x \in H_{p_1}^k, \quad y \in H_{p_2}^k \text{ and } H_p^k \cap (W_{1,k} \cup W_{2,k}) = \emptyset \text{ for all } p_1 < p < p_2,$$

then there exist $1 \leqslant q_1 < q_2 < q_3 < q_4 \leqslant n_k$ with $[\mathcal{G}_k(q_1, q_4)] \subset [\mathcal{H}_k(p_1, p_2)]$ and

$$x \in G_{q_1}^k = H_{p_1}^k, \ y \in G_{q_4}^k = H_{p_2}^k, \ G_{q_2}^k \subset H_{p_2-1}^k, \ G_{q_3}^k \subset H_{p_1+1}^k. \qquad (6.6)$$

Theorem 6.25 *For sequences $\{\mathcal{G}_k\}$, $k \in \mathbb{N}$, satisfying (A) and (B) above, $Q := \bigcap_{k \in \mathbb{N}}[\mathcal{G}_k]$ is a hereditarily indecomposable continuum and $\{a, b\} \subset Q$.*

Proof Since (A) holds for all $k \in \mathbb{N}$, $\{a, b\} \subset Q$ and by Theorem 3.43 $Q = \bigcap_{k \in \mathbb{N}}[\mathcal{G}_k]$ is a non-degenerate continuum.

Now if Q is not a hereditarily indecomposable continuum there exists a decomposable continuum $L \subset Q$ with $L = E \cup F$ where E and F are proper sub-continua of L. Therefore, there exist $e \in E \setminus F$, $f \in F \setminus E$ and $\ell \in \mathbb{N}$ such that

$$\text{dist}\,(e, F) > \frac{5}{\ell} \quad \text{and} \quad \text{dist}\,(f, E) > \frac{5}{\ell}. \qquad (6.7)$$

The remaining steps of this proof are illustrated by Fig. 6.2.

Now since $\mathcal{W}$ is a base for $\mathbb{R}^2$ there exists $k \geqslant \ell$ and $\{W_{1,k}, W_{2,k}\} \in \mathfrak{W}$ with

$$(e, f) \in W_{1,k} \times W_{2,k} \subset D_{1/\ell}(e) \times D_{1/\ell}(f). \qquad (6.8)$$

For this k let $\mathcal{H}_k = \{H_1^k, \cdots, H_{n_k'}^k\}$ and $\mathcal{G}_k = \{G_1^k, \cdots, G_{n_k}^k\}$ be the disc chains that satisfy (A) and (B), and such that the links of $\mathcal{H}_k$ are in $\boldsymbol{D}_{\epsilon_k}$. Then, since $e, f \in Q$, for this k there exist $\{p_1', p_2'\} \in \{1, \cdots, n_k'\}$ with $e \in H_{p_1'}^k$ and $f \in H_{p_2'}^k$, and so $|p_1' - p_2'| \geqslant 3$ by (6.4). Suppose without loss of generality that $p_1' < p_1' + 3 \leqslant p_2'$.

Since $e \in W_{1,k} \cap H_{p_1'}^k$ and $f \in W_{2,k} \cap H_{p_2'}^k$, there exist $p_1, p_2 \in \{p_1', \cdots, p_2'\}$, $p_1 < p_2$ and $(x, y) \in W_{1,k} \times W_{2,k}$ with

$$x \in H_{p_1}^k, \quad y \in H_{p_2}^k \text{ and } H_p^k \cap (W_{1,k} \cup W_{2,k}) = \emptyset \text{ for all } p_1 < p < p_2.$$

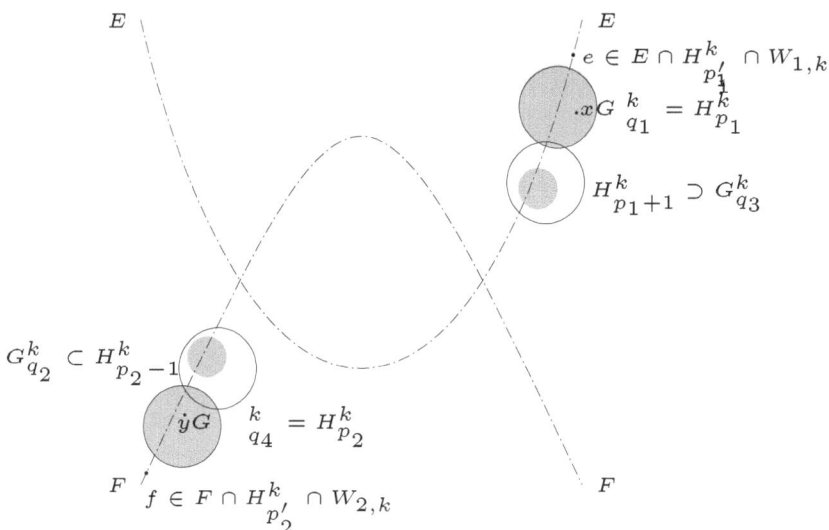

Fig. 6.2 This is a crude illustration of the construction used in the proof of Theorem 6.25. Here $L = E \cup F$ where $L \subset Q$ is a continuum in the plane, and E and F are sub-continua of L represented by parabola which are dashed because their internal structure is immensely more complicated. The black circles represent links of the disc chain $\mathscr{H}_k$ for some $k \geqslant \ell$, and the grey discs are the links of the sub-chain $\mathscr{G}_k(q_1, q_4)$ of $\mathscr{G}_k$ for $k \geqslant \ell$ in (6.7) and (6.8), which leads to the contradiction that F is not connected

Hence by (B) there exist $1 \leqslant q_1 < q_2 < q_3 < q_4 \leqslant n_k$ with

$$x \in G^k_{q_1} = H^k_{p_1}, \quad y \in G^k_{q_4} = H^k_{p_2}, \quad G^k_{q_2} \subset H^k_{p_2-1}, \quad G^k_{q_3} \subset H^k_{p_1+1}.$$

Now, since $\{e, f\} \subset L \subset [\mathscr{G}_k] \subset [\mathscr{H}_k]$, $e \in H^k_{p'_1}$ and $f \in H^k_{p'_2}$, and L is connected, it follows from Lemma 3.32 that all links of the sub-chain $\mathscr{H}_k(p'_1, p'_2)$ intersect L. Therefore, since

$$G^k_{q_1} \cap L = H^k_{p_1} \cap L \neq \emptyset \text{ and } G^k_{q_4} \cap L = H^k_{p_2} \cap L \neq \emptyset,$$

by the same argument all the links of $\mathscr{G}_k(q_1, q_4)$ intersect L. Moreover, by (6.3) $\text{diam}(H^k_p) \leqslant 2\epsilon_k \leqslant 2/k \leqslant 2/\ell$ for all p and hence, in the notation of (3.1a), $y \in W_{2,k} \cap H^k_{p_2} \subset H^k_{p_2} \subset N_{2/\ell}(W_{2,k}) \subset D_{3/\ell}(f)$ by (6.8), and so, by (6.7),

$$E \cap H^k_{p_2} = \emptyset. \tag{6.9}$$

Now by (6.6) and (6.8),

$$G^k_{q_2} \subset H^k_{p_2-1} \subset N_{2/\ell}(H^k_{p_2}) \subset N_{4/\ell}(W_{2,k}) \subset D_{5/\ell}(f),$$

$$G_{q_3}^k \subset H_{p_1+1}^k \subset N_{2/\ell}(H_{p_1}^k) \subset N_{4/\ell}(W_{1,k}) \subset D_{5/\ell}(e),$$

and hence, by (6.7) and (6.9),

$$E \cap G_{q_2}^k = \emptyset, \quad F \cap G_{q_3}^k = \emptyset \text{ and } E \cap G_{q_4}^k = \emptyset.$$

Since the links $G_{q_2}^k$ and $G_{q_4}^k$ intersect $L = E \cup F$, this shows that

$$F \cap G_{q_2}^k \neq \emptyset, \quad F \cap G_{q_4}^k \neq \emptyset \text{ and } F \cap G_{q_3}^k = \emptyset, \quad q_2 < q_3 < q_4,$$

and hence, by Lemma 3.32, that F is not connected.

But by hypothesis F is a continuum, a contradiction which shows that Q has no decomposable sub-continua, and the proof is complete.   □

6.5 Baire Category of Hereditarily Indecomposable Continua

The goal of this section is to show that $\mathcal{H}$, the set of all non-degenerate hereditarily indecomposable continua in $\mathbb{R}^2$, is of second Baire category (Definition 3.8) in the complete metric space $(C(\mathbb{R}^2), d_H)$ of all continua in $\mathbb{R}^2$ with the Hausdorff metric (Sect. 3.9 and [22, 23, 35, 41]).

Since $(C(\mathbb{R}^2), d_H)$ coincides with $(C(M), d_H)$ in Lemma 3.47 when $M = \mathbb{R}^2$, it follows from Lemmas 3.46 and 3.47 that $(C(\mathbb{R}^2), d_H)$ is a complete metric space, and therefore of second category by Baire's Category Theorem 3.12.

Therefore, to show that $\mathcal{H}$ is of second category in $(C(\mathbb{R}^2), d_H)$, by Lemma 3.10 it suffices to show that $\mathcal{H}$ contains the intersection of a countable family of sets each of which is open and dense in $(C(\mathbb{R}^2), d_H)$.

To begin the proof let a set of triples be defined by

$$\mathcal{T} = \Big\{ (a, b, \mathcal{G}) : a \neq b \in \mathbb{R}^2 \text{ and } \mathcal{G} = \{G_1, \cdots, G_n\} \text{ is a}$$

$$\text{strict disc chain joining } a \text{ to } b \text{ with } a \notin \overline{G}_2 \text{ and } b \notin \overline{G}_{n-1} \Big\}. \quad (6.10)$$

Note from Definitions 3.29 and 6.15 that if $(a, b, \mathcal{G}) \in \mathcal{T}$, then $a \in G_1$ but $a \notin \overline{G}_j$, $j \in \{2, \cdots, n\}$ and $b \in G_n$ but $b \notin \overline{G}_j$, $j \in \{1, \cdots, n-1\}$, and $n \geqslant 2$. Also note that, in (6.10), the conditions $a \notin \overline{G}_2$ and $b \notin \overline{G}_{n-1}$ will be satisfied by all strict disc chains that follow because they are obtained with the help of Lemma 6.21.

Now, for a countable base $\mathscr{W}$ in $\mathbb{R}^2$, let $\mathfrak{W}$ and $\{\epsilon_k\}_{k \in \mathbb{N}}$ be as defined by (6.2) and (6.3), and for any $(a, b, \mathcal{G}_0) \in \mathcal{T}$ let $\{\mathcal{G}_k\}_{k \geqslant 0}$ be a sequence of strict disc chains starting with $\mathcal{G}_0$, constructed recursively as in (A) and (B) of the preceding section; in particular

$$(a, b, \mathcal{G}_k) \in \mathcal{T} \quad \text{for all } k \in \mathbb{N}.$$

Although $\mathcal{G}_k$, $k > 1$, is not uniquely determined by $T := (a, b, \mathcal{G}_0) \in \mathcal{T}$, let the dependence on T of a sequence of disc chains generated by (A) and (B) be denoted by $\{\mathcal{G}_k(T)\}_{k \geqslant 0}$, where each $\mathcal{G}_k(T)$ has $n_k = n_k(T) \geqslant 2$ links.

Then the corresponding sequences of strict disc chains $\{\mathcal{G}_k(T)\}_{k \geqslant 0}$ and $\{\mathcal{H}_k(T)\}_{k \geqslant 1}$ joining a to b which satisfy (A) and (B) will be denoted by

$$\mathcal{G}_k(T) = \left(G_1^k(T), \ldots, G_{n_k(T)}^k(T)\right), \quad n_k(T) \geqslant 2, \quad k \geqslant 0, \tag{6.11}$$
$$\mathcal{H}_k(T) = \left(H_1^k(T), \ldots, H_{n_k'(T)}^k(T)\right), \quad n_k'(T) \geqslant 2, \quad k \geqslant 1,$$

the links of $\mathcal{H}_k(T)$ being in $\mathbf{D}_{\epsilon_k}$. Henceforth, for given $T \in \mathcal{T}$ the sequences $\{\mathcal{G}_k(T)\}_{k \geqslant 0}$ and $\{\mathcal{H}_k(T)\}_{k \geqslant 1}$ in (6.11) are chosen arbitrarily, but then fixed.

With this understanding Theorem 6.25 can be generalised as follows.

For each integer $\ell \geqslant 0$ let $T_\ell = (a_\ell, b_\ell, \mathcal{G}_{0,\ell}) \in \mathcal{T}$ and let $\{\mathcal{G}_k(T_\ell)\}_{k \geqslant 0}$ with $\mathcal{G}_0(T_\ell) = \mathcal{G}_{0,\ell}$ and $\{\mathcal{H}_k(T_\ell)\}_{k \geqslant 1}$ be sequences satisfying (A), (B) and (6.11). Then let $\{\widehat{\mathcal{G}}_k\}$ be the diagonal sequence $\{\mathcal{G}_k(T_k)\}$ and let

$$\widehat{Q} := \bigcap_{k \geqslant 1} [\widehat{\mathcal{G}}_k] = \bigcap_{k \geqslant 1} [\mathcal{G}_k(T_k)].$$

Note that $\{[\widehat{\mathcal{G}}_k]\}$ may not be nested and $\widehat{Q}$ may be empty or not connected. Nevertheless, the following is a generalisation of Theorem 6.25.

Theorem 6.26 *If $Q \subset \widehat{Q}$ is a non-degenerate continuum, $Q \in \mathcal{H}$.*

Proof The proof is essentially the same as that of Theorem 6.25. Indeed, in the proof of Theorem 6.25, a, b, $\mathcal{G}_0$ are not used, and only a carefully chosen value of k is considered explicitly. □

Now for any strict disc chain $\mathcal{G} = \{G_1, \ldots, G_n\}$, let $C(\mathcal{G})$ denote the set of non-degenerate continua $C \subset [\mathcal{G}]$ that intersect $G_1 \backslash \overline{G_2}$ and $G_n \backslash \overline{G_{n-1}}$, and hence $C \in C(\mathcal{G})$ intersects every link of $\mathcal{G}$ by Lemma 3.32.

Corollary 6.27 $\bigcap_{k \geqslant 1} \left(\bigcup_{T \in \mathcal{T}} C(\mathcal{G}_k(T)) \right)$ *is a subset of $\mathcal{H}$, the set of all non-degenerate hereditarily indecomposable continua in $\mathbb{R}^2$.*

Proof If a non-degenerate continuum Q belongs to $\bigcap_{k \geqslant 1} \bigcup_{T \in \mathcal{T}} C(\mathcal{G}_k(T))$, then for all $k \in \mathbb{N}$ there exists $T_k \in \mathcal{T}$ such that $Q \in C(\mathcal{G}_k(T_k))$. Hence $Q \subset \widehat{Q} = \bigcap_{k \in \mathbb{N}} [\mathcal{G}_k(T_k)]$ and the result follows from Theorem 6.26. □

Therefore, to show that $\mathcal{H}$ is of second Baire category in $(C(\mathbb{R}^2), d_H)$ it suffices, by Lemma 3.10 and Corollary 6.27, to prove the following.

Theorem 6.28 *The set $\bigcup_{T \in \mathcal{T}} C(\mathcal{G}_k(T))$ is open and dense in $(C(\mathbb{R}^2), d_H)$ for all $k \in \mathbb{N}$.*

Proof Since $[\mathscr{G}] \subset \mathbb{R}^2$ is open for any strict disc chain $\mathscr{G}$, it follows that $C(\mathscr{G})$ is open in $\left(C(\mathbb{R}^2), d_H \right)$ and hence $\bigcup_{T \in \mathscr{T}} C(\mathscr{G}_k(T))$ is open in $\left(C(\mathbb{R}^2), d_H \right)$.

To see that it is dense, let a continuum $C \subset \mathbb{R}^2$ and $\delta > 0$ be arbitrary. Then there is a finite set, $D_{\delta/4}(a_1), \ldots, D_{\delta/4}(a_m)$, of open discs each of which meets C, with distinct centres $a_1, \ldots, a_m$, $m \geqslant 2$, and $C \subset A := \bigcup_{j=1}^m D_{\delta/4}(a_j)$. Now suppose that for the set $\{a_1, \cdots, a_m\}$ there is a strict disc chain $\mathscr{G}$ with

$$\{a_1, \cdots, a_m\} \subset [\mathscr{G}] \subset A, \ (a_1, a_m, \mathscr{G}) \in \mathscr{T} \text{ and links in } D_{\delta/4}. \tag{6.12}$$

Then for such a $\mathscr{G}$, by Lemma 4.2 there is a path $P \subset [\mathscr{G}]$ joining a_1 to a_m and since P is connected and a_1 and a_m are in the end links of $\mathscr{G}$, P intersects all the links of $\mathscr{G}$. Therefore

$$\{a_1, \ldots, a_m\} \subset [\mathscr{G}] \subset N_{\delta/2}(P), \ P \subset A = N_{\delta/4}(\{a_1, \ldots, a_m\})$$

and $d_H(\{a, \ldots, a_m\}, P) \leqslant \delta/2$. Moreover

$$\{a_1, \ldots, a_m\} \subset N_{\delta/4}(C), \ C \subset A = N_{\delta/4}(\{a_1, \ldots, a_m\})$$

and $d_H(C, \{a, \ldots, a_m\}) \leqslant \delta/4$. Hence

$$d_H(C, P) \leqslant d_H(C, \{a_1, \ldots, a_m\}) + d_H(\{a_1, \ldots, a_m\}, P) \leqslant \frac{\delta}{4} + \frac{\delta}{2} < \delta. \tag{6.13}$$

In particular, if $T_0 := (a_1, a_m, \mathscr{G}) \in \mathscr{T}$ and P_k is a path in $[\mathscr{G}_k(T_0)] \subset [\mathscr{G}]$ joining a_1 to a_m, it follows that $d_H(C, P_k) < \delta$ by (6.13). Hence, for each $k \in \mathbb{N}$, there are continua in the open set $\bigcup_{T \in \mathscr{T}} C(\mathscr{G}_k(T))$ arbitrarily close to C in $\left(C(\mathbb{R}^2), d_H \right)$. Thus, (6.12) implies that $\bigcup_{T \in \mathscr{T}} C(\mathscr{G}_k(T))$ is open and dense in $(C(\mathbb{R}^2), d_H)$ for any $k \in \mathbb{N}$, as required.

It remains only to show the existence of $\mathscr{G}$ satisfying (6.12) for the given set $\{a_1, \cdots, a_m\}$, $m \geqslant 2$. So choose $r \in (0, \delta/4)$ small enough that all the closed discs $\overline{D_r(a_j)}$, $1 \leqslant j \leqslant m$, are mutually disjoint subsets of $A = \bigcup_{j=1}^m D_{\delta/4}(a_j)$ and, by Lemma 6.18, $A \setminus \bigcup_{j>2} \overline{D_r(a_j)}$ (A if $m = 2$) is connected. Then by Lemma 6.21, for $\epsilon \in (0, r)$ there is a strict disc chain $\widetilde{\mathscr{D}} = \{\widetilde{D}_1, \cdots, \widetilde{D}_n\} \subset D_\epsilon$ in $A \setminus \bigcup_{j>2} \overline{D_r(a_j)}$ with $\widetilde{D}_1 = D_{\epsilon/2}(a_1)$, $\widetilde{D}_n = D_{\epsilon/2}(a_2)$, joining a_1 to a_2.

By Lemma 6.19, $\epsilon > 0$ and $\widetilde{\mathscr{D}}$ can be chosen so that $[\widetilde{\mathscr{D}}] \subset A \setminus \left(\bigcup_{j>2} \overline{D_r(a_j)} \right)$ and $A \setminus \left\{ [\widetilde{\mathscr{D}}(1, n-1)] \cup \left(\bigcup_{j>2} \overline{D_r(a_j)} \right) \right\}$ is open and connected. Then if $m = 2$, put $\mathscr{G} = \widetilde{\mathscr{D}}$ to satisfy (6.12).

Then to proceed recursively suppose $m' \in \{3, \ldots, m\}$ is such that for some $\epsilon \in (0, r)$ there is a strict disc chain $\widetilde{\mathscr{D}} = \{\widetilde{D}_1, \cdots, \widetilde{D}_n\} \subset D_\epsilon$ in $A \setminus \left(\bigcup_{j \geqslant m'} \overline{D_r(a_j)} \right)$ with $\widetilde{D}_1 = D_{\epsilon/2}(a_1)$, $\widetilde{D}_n = D_{\epsilon/2}(a_{m'-1})$, which joins a_1 to a_2 if $m' = 3$, and joins a_1 to $a_{m'-1}$ via $\{a_2, \ldots, a_{m'-2}\}$, if $m' > 3$. Moreover, by Lemma 6.19 suppose $A \backslash \left\{ \overline{[\widetilde{\mathscr{D}}(1, n-1)]} \cup \left(\bigcup_{j \geqslant m'} \overline{D_r(a_j)} \right) \right\}$ is open and connected and $\overline{[\widetilde{\mathscr{D}}]} \subset A \backslash \left(\bigcup_{j \geqslant m'} \overline{D_r(a_j)} \right)$.

Since $\widehat{A} := \widetilde{D}_n \cup \left\{ A \setminus \left(\overline{[\widetilde{\mathscr{D}}(1, n-1)]} \cup \left(\bigcup_{j > m'} \overline{D_r(a_j)} \right) \right) \right\}$ is open and connected, and $D_{\epsilon/2}(a_{m'-1}) \cup D_{\epsilon/2}(a_{m'}) \subset \widehat{A}$, by Lemma 6.21 there is a strict disc chain $\widehat{\mathscr{D}} = \{\widehat{D}_1, \cdots, \widehat{D}_{\widehat{n}}\} \subset \boldsymbol{D}_\epsilon$ in $\widehat{A}$ joining $a_{m'-1}$ to $a_{m'}$, and $\widehat{D}_1 = \widetilde{D}_n$ with $\widehat{D}_{\widehat{n}} = D_{\epsilon/2}(a_{m'})$. Therefore $\mathscr{D} = \{\widetilde{D}_1, \ldots, \widetilde{D}_{n-1}, \widetilde{D}_n = \widehat{D}_1, \widehat{D}_2, \ldots, \widehat{D}_{\widehat{n}}\}$ is a strict disc chain with $n + \widehat{n} - 1$ links in $\boldsymbol{D}_\epsilon$ joining a_1 to $a_{m'}$ via $a_2, \ldots, a_{m'-1}$,

$$[\mathscr{D}] \subset A \setminus \bigcup_{j > m'} \overline{D_r(a_j)}$$

and the only link containing $a_{m'-1}$ is $\widetilde{D}_n = \widehat{D}_1$. By Lemma 6.19, $\epsilon > 0$ and $\mathscr{D}$ can be chosen such that $A \setminus \left\{ \overline{[\mathscr{D}(1, n-1)]} \cup \left(\bigcup_{j > m'} \overline{D_r(a_j)} \right) \right\}$ is connected and $\overline{[\mathscr{D}]} \subset A \backslash \left(\bigcup_{j > m'} \overline{D_r(a_j)} \right)$.

Proceeding recursively until $m' = m$ is reached, a strict disc chain $\mathscr{G}$ satisfying (6.12) has been constructed, and the proof is complete. $\qquad\square$

Chapter 7
Pathological Examples

7.1 Examples of Global Bifurcation Without Paths

In this section a variety of examples of problem (1.1) will be constructed to show as in [48] that although Theorem 1.3 predicts the bifurcation of global connected sets of non-trivial solutions these sets may not be path-connected even when the function R is continuously differentiable everywhere and infinitely differentiable except at one point.

The setting for these examples is very basic: $X = \mathbb{R}$, $L : \mathbb{R} \to \mathbb{R}$ is the identity operator and $R : \mathbb{R}^2 \to \mathbb{R}$ is continuous and satisfies (1.1b). Thus Eq. (1.1a) has the form

$$x = \lambda x + R(\lambda, x), \quad (\lambda, x) \in \mathbb{R}^2, \tag{7.1}$$

and by the remark before Definition 1.2 the only possible bifurcation point is $\lambda_0 = 1$, and by Theorem 1.3 an unbounded connected set of non-trivial solutions of (7.1) bifurcates from $\mathcal{T}$ at $(1, 0) \in \mathbb{R}^2$.

Example A For a C^1-function $R : \mathbb{R}^2 \to \mathbb{R}$ which is infinitely differentiable on $\mathbb{R}^2 \setminus \{(1, 0)\}$, the components of non-trivial solutions of (7.1) are unbounded but contain no non-trivial paths, see Fig. 7.2. □

Example B In this example of (7.1), $R = \widetilde{R}$ where $\widetilde{R} : \mathbb{R}^2 \to \mathbb{R}$ is a C^∞-function, the closure of the set of non-trivial solutions of (7.1) is a connected set $\overline{\mathscr{C}} = \mathscr{C}^- \cup \mathscr{L} \cup \mathscr{C}^+$ where $\mathscr{C}^-$, $\mathscr{C}^+$ and $\mathscr{L}$ are disjoint, $\mathscr{L}$ is a smooth curve of solutions, $\mathscr{C}^\pm$ are closed, unbounded, connected sets and all the non-trivial paths in $\mathscr{C}^- \cup \mathscr{L} \cup \mathscr{C}^+$ are subsets of $\overline{\mathscr{L}}$. □

Example C In this example of (7.1), for a C^∞-function R the two possibilities in Sect. 5.4 (β)(ii) occur simultaneously, but at different points, $(1, \pm 1/2) \in (\partial_{\mathscr{M}} Q) \setminus \mathscr{N}(\mathscr{M})$ in Fig. 7.3. □

© The Author(s), under exclusive license to Springer Nature Switzerland AG 2025
B. Buffoni, J. Toland, *Connected Sets in Global Bifurcation Theory*, SpringerBriefs
in Mathematics, https://doi.org/10.1007/978-3-031-87051-4_7

These examples are special cases of the following trivial observation.

Theorem 7.1 *Suppose $F \subset \mathbb{R}^2$ is closed and $f : \mathbb{R}^2 \to \mathbb{R}$ is continuous with*

$$f(\lambda, 0) = \lambda \text{ for } \lambda \in \mathbb{R} \text{ and } f(\lambda, x) = 0 \text{ if and only if } (\lambda, x) \in F. \tag{7.2}$$

Then

$$R(\lambda, x) = x\big(f(\lambda - 1, x) - f(\lambda - 1, 0)\big) \tag{7.3}$$

defines a continuous function $R : \mathbb{R}^2 \to \mathbb{R}$ which satisfies (1.1b) and (λ, x) is a non-trivial solution of (7.1) if and only if $(\lambda, x) \in F + (1, 0)$ and $x \neq 0$.

Proof By (7.3), R satisfies (1.1b) since f is continuous on $\mathbb{R}^2$. Moreover by (7.2), for $(\lambda, x) \in \mathbb{R}^2$,

$$x - \lambda x - R(\lambda, x) = x - \lambda x - x f(\lambda - 1, x) + x f(\lambda - 1, 0)$$

$$= x - \lambda x - x f(\lambda - 1, x) + x(\lambda - 1) = -x f(\lambda - 1, x).$$

Hence (λ, x) satisfies (7.1) if and only if $(\lambda - 1, x) \in F$ or $x = 0$, that is either $(\lambda, x) \in (1, 0) + F$ or $x = 0$. The result for non-trivial solutions follows. □

Therefore, for a set F with prescribed properties, the aim is to find a function f as in Theorem 7.1 such that R satisfies the hypotheses of Theorem 1.3.

An Example

For a fixed $p \in \mathbb{N}$ with $2k < p$, $k \in \mathbb{N}$, let f in Theorem 7.1 be given by

$$f(\lambda, x) = (1 - x^2)^p \left\{ \lambda + \sin\left(\frac{x}{1 - x^2}\right) \right\}, \quad x \in \mathbb{R} \setminus \{1, -1\},$$

$$f(\lambda, \pm 1) = 0, \quad \lambda \in \mathbb{R}.$$

Then $f \in C^k$, $f(\lambda, 0) = \lambda$ for all $\lambda \in \mathbb{R}$, and the zero set $F = \{(\lambda, x) \in \mathbb{R}^2 : f(\lambda, x) = 0\}$ is

$$F = \left\{ (\lambda, \pm 1) : \lambda \in \mathbb{R} \right\} \bigcup \left\{ (\lambda, x) \in \mathbb{R}^2 : \lambda + \sin\left(\frac{x}{1 - x^2}\right) = 0, \quad x \neq \pm 1 \right\}.$$

Note that F contains the union of four curves similar to the topologist's sine curve (Example 4.5) and its set of congestion points has two path-components, namely $[-1, 1] \times \{1\}$ and $[-1, 1] \times \{-1\}$. Therefore the solution set, Fig. 7.1, of equation

Fig. 7.1 Global bifurcation of a generalised continuum in $\mathbb{R}^2$ which has five path-components and a set of congestion points which has two path-components

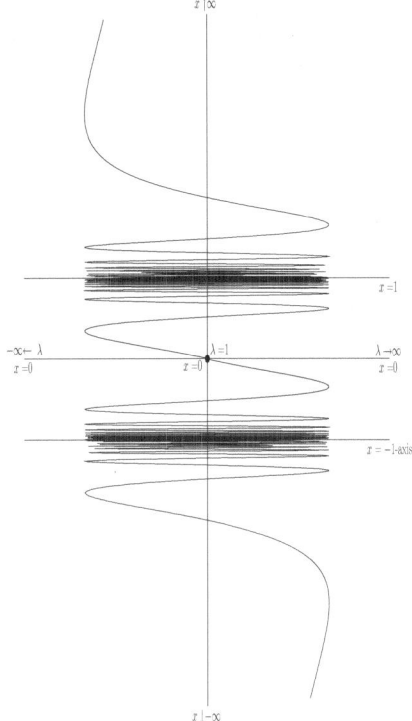

(7.1) with

$$R(\lambda, x) = \begin{cases} x\left((1 - x^2)^p \left\{\lambda - 1 + \sin\left(\frac{x}{1-x^2}\right)\right\} + 1 - \lambda\right), & |x| \neq 1, \\ \operatorname{sign}(x)(1 - \lambda), & |x| = 1, \end{cases}$$

has an unbounded bifurcating connected set which is not path-connected. □

The constructions of Examples A, B and C depend on the next result which seems to be well known but is difficult to pin down explicitly, see [45, Ch. 6, §2] for related results.

Lemma 7.2 *For a closed set $F \subset \mathbb{R}^2$ there is a function $h : \mathbb{R}^2 \to \mathbb{R}$ which is non-negative, infinitely differentiable and globally Lipschitz continuous such that $F = \{x \in \mathbb{R}^2 : h(x) = 0\}$, and all the derivatives of h are zero on F.*

Remark The function h is globally Lipschitz continuous if there is a constant K such that $|h(x) - h(y)| \leqslant K\|x - y\|$ for all $x, y \in \mathbb{R}^2$. □

Proof Let $u : [0, \infty) \to [0, 1]$ be a C^∞-function with

$$u(t) = 1, \quad t \in [0, 1/2]; \quad u(t) \in (0, 1), \quad t \in (1/2, 1); \quad u(t) = 0, \quad t \in [1, \infty).$$

For the closed set F, let the open set $\mathbb{R}^2 \setminus F$ be the union of a countable collection of open balls $\{B_{r_j}(a_j) : j \in \mathbb{N}\}$, with radius $r_j \in (0, 1)$ centred at $a_j \in \mathbb{R}^2$, and put

$$u_j(x) = u\left(\frac{|x - a_j|}{r_j}\right), \quad x \in \mathbb{R}^2.$$

Then $u_j \geqslant 0$ is infinitely differentiable on $\mathbb{R}^2$ and positive on $B_{r_j}(a_j)$. Now for $k \in \mathbb{N} \cup \{0\}$ let $\mathrm{L}^k(\mathbb{R}^2, \mathbb{R})$ denote the linear space of k-linear maps $(\mathbb{R}^2)^k \to \mathbb{R}$ and let $\|A\|_k$ denote the norm of $A \in \mathrm{L}^k(\mathbb{R}^2, \mathbb{R})$ (see e.g. [7]). Then since, for each $j \in \mathbb{N}$ and $k \in \mathbb{N} \cup \{0\}$, $D^k u_j(x) \in \mathrm{L}^k(\mathbb{R}^2, \mathbb{R})$ is the kth derivative of u_j at $x \in \mathbb{R}^2$ and $D^0 u_j(x) = u_j(x)$ it follows that $D^k u_j(x) = 0$ for all $x \in F$ and $k \in \mathbb{N} \cup \{0\}$. Moreover, since u_j is supported on $\overline{B_{r_j}(a_j)}$,

$$0 < \gamma_j := \max\{\|D^k u_j(x)\|_k : 0 \leqslant k \leqslant j, x \in \mathbb{R}^2\} < \infty \text{ for all } j \in \mathbb{N}.$$

It follows that $\gamma_j \leqslant \gamma_{j+1}$,

$$\sum_{j=1}^{\infty} \frac{\|u_j(x)\|}{\gamma_j\, 2^j} \leqslant \sum_{j=1}^{\infty} \frac{1}{2^j} = 1,$$

and for all $k \in \mathbb{N}$

$$\sum_{j=k}^{\infty} \frac{\|D^k u_j(x)\|}{\gamma_j\, 2^j} \leqslant \sum_{j=k}^{\infty} \frac{1}{2^j} = \frac{1}{2^{k-1}},$$

from which it follows that, for all $x \in \mathbb{R}^2$,

$$h(x) = \sum_{j \in \mathbb{N}} \frac{u_j(x)}{\gamma_j\, 2^j} \in \mathbb{R} \text{ is well defined}$$

and, for all $k \in \mathbb{N}$,

$$\sum_{j=1}^{k-1} \frac{D^k u_j(x)}{\gamma_j\, 2^j} + \sum_{j=k}^{\infty} \frac{D^k u_j(x)}{\gamma_j\, 2^j} \in \mathrm{L}^k(\mathbb{R}^2, \mathbb{R}),$$

where the series converges uniformly in x. Therefore for $k \in \{0\} \cup \mathbb{N}$,

$$D^k h(x) = \sum_{j \in \mathbb{N}} \frac{D^k u_j(x)}{\gamma_j\, 2^j} \in \mathrm{L}^k(\mathbb{R}^2, \mathbb{R}).$$

and in particular

$$\|D^1 h(x)\| \leqslant \sum_{j \in \mathbb{N}} \frac{\|D^1 u_j(x)\|}{\gamma_j \, 2^j} \leqslant \sum_{j \in \mathbb{N}} \frac{1}{2^j} = 1.$$

Hence h is globally Lipschitz continuous and infinitely differentiable, and since $D^k u_j(x) = 0$ for all $x \in F$, $k \geqslant 0$ and $j \geqslant 1$, $F = \{x \in \mathbb{R}^2 : h(x) = 0\}$ and $D^k h(x) = 0$ for $k \in \mathbb{N}$ and $x \in F$. This completes the proof.   □

7.2 Preliminaries

The following criterion is to ensure that certain connected sets separate the plane. Let K be the double cone centred on the vertical x-axis defined by

$$K := \{(\lambda, x) \in \mathbb{R}^2 : |\lambda| < |x|\} \cup \{(0, 0)\}. \tag{7.4}$$

Definition 7.3 A set $F \subset \mathbb{R}^2$ satisfies (†) if it is closed, connected, $F \subset K$ and there are sequences

$$x_j^\pm \to \pm\infty \text{ as } j \to \infty \text{ with } F \cap \left(\mathbb{R} \times \{x_j^\pm\}\right) \neq \emptyset.   □$$

Lemma 7.4 *Suppose F satisfies (†). Then $(0, 0) \in F$ and $H^+ \cap H^- = \emptyset$ where $H^\pm$ are the components of $\mathbb{R}^2 \setminus F$ with $(\pm 1, 0) \in H^\pm$.*

Remark $\mathbb{R}^2 \setminus F$ may have other components.   □

Proof Note that $(0, 0) \in F$ since otherwise F is separated by $\{(\lambda, x) \in F, x > 0\}$ and $\{(\lambda, x) \in F, x < 0\}$, and this is false. Note also that if $(1, 0)$ and $(-1, 0)$ are in the same component, $H = H^+ = H^-$ of $\mathbb{R}^2 \setminus F$ say, since $\mathbb{R}^2 \setminus F$ is open, H is an open connected set in $\mathbb{R}^2$ by Corollary 4.4.

Now let $\mathscr{D}$ be a strict disc chain in H joining $(-1, 0)$ to $(1, 0)$, the centres of the end discs being $(-1, 0)$ and $(1, 0)$ (see Lemma 6.21). Joining each pair of consecutive centres by straight lines yields a piecewise-affine path Γ in H, which is non-self-intersecting, with finitely many straight-line segments, and joining $(-1, 0)$ and $(1, 0)$.   □

By slightly modifying the centres of the disc-chain links, it can be assumed that Γ has no vertical and no horizontal segments. Since $(0, 0) \notin \Gamma$ because $(0, 0) \in F$, there is a non-self-intersecting polygonal sub-path $\gamma = \{(\lambda(t), x(t)) : t \in [0, 1]\}$ of Γ with $\{(\lambda(t), x(t)) : t \in (0, 1)\} \subset K \setminus \{(0, 0)\}$, $\{(\lambda(0), x(0)), (\lambda(1), x(1))\} \subset (\partial K) \setminus \{(0, 0)\}$ and $\lambda(0) < 0 < \lambda(1)$.

Now let $P \subset \mathbb{R}^2 \backslash F$ denote the union of the sub-path γ with $(-\infty, \lambda(0)] \times \{x(0)\}$ and $[\lambda(1), \infty) \times \{x(1)\}$, so that the segments of P are these two horizontal semi-infinite half-lines and the segments of γ, none of which is vertical.

Then, for any $A = (\lambda, x) \in \mathbb{R}^2 \backslash P$, let $V_A = \{\lambda\} \times [x, \infty)$ be the semi-infinite vertical half-line with lower endpoint at A. Then, for $A \in \mathbb{R}^2 \setminus P$, points of intersection of P with V_A, if any, are isolated and there are four possibilities:

(a) $V_A \cap P = \emptyset$;
(b) V_A meets P at a point which is not a vertex of P;
(c) V_A meets P at a vertex of P and the two adjacent segments of P (finite or semi-infinite) are on opposite sides (horizontally) of V_A;
(d) V_A meets P at a vertex of P and the two adjacent segments of P are on the same side of V_A.

Now let

$$\rho(A) = (-1)^{\nu(A)}, \quad A \in \mathbb{R}^2 \setminus P,$$

where $\nu(A)$ is the count of all points in categories (a), (b) and (c).

Remark $\rho(A)$ is a mod-2 count of the non-degenerate crossings of P by V_A for which category (d) crossings are degenerate.   □

Since $\rho(A) = 1$ if (a) holds or if the number of crossings in categories (b) or (c) is even, and $\rho(A) = -1$ if the number of category (b) or (c) crossings is odd, the function $\rho : \mathbb{R}^2 \backslash P \to \{-1, 1\}$ is locally constant. Now since the sub-path γ is bounded and ρ changes sign when A crosses one of the two horizontal lines $(-\infty, \lambda(0)) \times \{x(0)\}$ and $(\lambda(1), \infty) \times \{x(1)\}$, it follows that $\rho^{-1}(\{1\})$ and $\rho^{-1}(\{-1\})$ are open and

$$\mathbb{R} \times \{\pm x\} \subset \rho^{-1}(\{\pm 1\}) \text{ for } x > 0 \text{ sufficiently large.}$$

On the other hand, by Definition 7.3 there exist two sequences $x_j^\pm \to \pm\infty$ as $j \to \infty$ with $F \cap (\mathbb{R} \times \{x_j^\pm\}) \neq \emptyset$. Choose $A_j^\pm \in F \cap (\mathbb{R} \times \{x_j^\pm\})$, $j \in \mathbb{N}$. Then $\rho(A_j^\pm) = \pm 1$ for all j sufficiently large and hence F is separated by $\rho^{-1}(\{1\}) \cap F$ and $\rho^{-1}(\{-1\}) \cap F$. Since F is connected, this is a contradiction from which it follows that $H^+ \cap H^- = \emptyset$, where $H^\pm$ are the components of $\mathbb{R}^2 \setminus F$ with $(\pm 1, 0) \in H^\pm$.

Remark In the preceding proof the fact that P is a polygonal path with no vertical segments simplified the proof of what is basically a version of the Jordan curve theorem about separating the plane.   □

Lemma 7.5 *For K in (7.4) there exists $\omega : \mathbb{R}^2 \to \mathbb{R}$ with the following properties:*

(i) $\omega(\lambda, x) = 0$ *if* $|x| \geq |\lambda|/2$; *in particular* $\omega = 0$ *on* K;

(ii) $\lambda \omega(\lambda, x) \geqslant 0$ on $\mathbb{R}^2$;
(iii) $\omega(\lambda, 0) = \lambda$, $\quad \lambda \in \mathbb{R}$;
(iv) ω is infinitely differentiable on $\mathbb{R}^2 \setminus \{(0, 0)\}$;
(v) ω is globally Lipschitz continuous on $\mathbb{R}^2$.

Proof Let $\varphi : \mathbb{R} \to \mathbb{R}$ be an infinitely differentiable even function which is non-increasing on $[0, \infty)$ with $\varphi(0) = 1$ and $\varphi(r) = 0$ for all $r \geqslant 1/2$. Then, for $x \in \mathbb{R}$, let

$$\omega(\lambda, x) = \lambda \varphi \left(\frac{x}{\lambda} \right), \quad \lambda \neq 0, \quad \omega(0, x) = 0.$$

Clearly ω is continuous on $\mathbb{R}^2$ and odd in λ, and since it is zero in a neighbourhood of every point of the x-axis ($\lambda = 0$) except $(0, 0)$, properties (i)-(iv) follow from the properties of φ. Moreover, the partial derivatives of ω are

$$\partial_x \omega(\lambda, x) = \varphi' \left(\frac{x}{\lambda} \right), \quad \partial_\lambda \omega(\lambda, x) = \varphi \left(\frac{x}{\lambda} \right) - \left(\frac{x}{\lambda} \right) \varphi' \left(\frac{x}{\lambda} \right), \quad \lambda \neq 0, \tag{7.5a}$$

$$\partial_x \omega(0, x) = \partial_\lambda \omega(0, x) = 0 \text{ when } x \neq 0, \tag{7.5b}$$

$$\partial_x \omega(0, 0) = 0, \quad \partial_\lambda \omega(0, 0) = 1. \tag{7.5c}$$

Since $\varphi'(r) = 0$, $|r| \geqslant 1/2$, the partial derivatives of ω are uniformly bounded in $\mathbb{R}^2 \setminus \{(0, 0)\}$, and property (v) follows. $\qquad \square$

For future reference note that

$$\partial_{x\lambda} \big(x \, \omega(\lambda, x) \big) = \varphi \left(\frac{x}{\lambda} \right) - \left(\frac{x}{\lambda} \right) \varphi' \left(\frac{x}{\lambda} \right) - \left(\frac{x}{\lambda} \right)^2 \varphi'' \left(\frac{x}{\lambda} \right), \quad \lambda \neq 0. \tag{7.5d}$$

Remark 7.6 It follows from (7.5b) and (7.5c) that $\partial_\lambda \omega$ is not continuous at $(0, 0)$ and from (7.5a) and (7.5c) that $\partial_x \omega$ is not continuous at $(0, 0)$. However, $(\lambda, x) \mapsto x \, \omega(\lambda, x)$ is continuously differentiable on $\mathbb{R}^2$. Note also, from the intermediate value theorem, that for any $\rho \in (0, 1)$ there exists $s \in (0, 1/2)$ such that $\varphi(s) - s\varphi'(s) - s^2\varphi''(s) = \rho$ and hence, from (7.5d) with $x = s\lambda$, $\lambda \neq 0$, that

$$\partial_{x\lambda} \big(x\omega(\lambda, x) \big) \Big|_{(\lambda, s\lambda)} \to \rho \text{ as } \lambda \to 0.$$

Moreover

$$\partial_{\lambda x} \big(x\omega(\lambda, x) \big) \Big|_{(0,0)} = \partial_\lambda \omega(\lambda, 0) \big|_{\lambda=0} = \partial_\lambda \lambda \big|_{\lambda=0} = 1$$

and

$$\partial_{x\lambda} \big(x\omega(\lambda, x) \big) \Big|_{(0,0)} = \partial_x \big(x \partial_\lambda \omega(0, x) \big) \big|_{x=0} = \partial_x (x \cdot 0) \big|_{x=0} = 0,$$

and thus both mixed partial derivatives $\partial_{\lambda x}\big(x\,\omega(\lambda, x)\big)$ and $\partial_{x\lambda}\big(x\,\omega(\lambda, x)\big)$ exist but are not continuous at $(0, 0)$.   □

Theorem 7.7 *Suppose F satisfies* (†) *in Definition 7.3. Then $(\lambda, 0) \in F$ if and only if $\lambda = 0$ and there is a locally Lipschitz continuous function $f : \mathbb{R}^2 \to \mathbb{R}$ which is infinitely differentiable on $\mathbb{R}^2 \setminus \{(0, 0)\}$ and for $\lambda \in \mathbb{R}$,*

$$f(\lambda, 0) = \lambda \text{ and when } x \neq 0, \ \ f(\lambda, x) = 0 \text{ if and only if } (\lambda, x) \in F.$$

Proof Since $K \setminus \{(0, 0)\}$ is not connected but F is, it follows that $(0, 0) \in F$, and $(0, 0)$ is the only point $(\lambda, 0) \in F$ since $F \subset K$. Also, by Lemma 7.4, $H^+ \cap H^- = \emptyset$ where $H^\pm$ are the components of $\mathbb{R}^2 \setminus F$ with $(\pm 1, 0) \in H^\pm$. Moreover, since F is closed, by Lemma 7.2 there is a non-negative, infinitely differentiable function $h : \mathbb{R}^2 \to [0, \infty)$ such that $h(\lambda, x) = 0$ if and only if $(\lambda, x) \in F$, and all the derivatives of h are zero at every point of F.

Let $\hat{h} : \mathbb{R}^2 \to \mathbb{R}$ be defined by

$$\hat{h}(\lambda, x) = \begin{cases} -h(\lambda, x), & (\lambda, x) \in H^-, \\ h(\lambda, x), & \text{otherwise,} \end{cases}$$

so that $\hat{h}(\lambda, x) = \pm h(\lambda, x)$ when $(\lambda, x) \in H^\pm$ and $\hat{h}(\lambda, x) = 0$ if and only if $(\lambda, x) \in F$. To see that $\hat{h}$ is infinitely differentiable, as in the proof of Lemma 7.2 let the open set $\mathbb{R}^2 \setminus F$ be the union of countably open balls $\{B_{r_j}(a_j) : j \in \mathbb{N}\}$, centred at $a_j \in \mathbb{R}^2$. Then with $J^- = \{j \in \mathbb{N} : a_j \in H^-\}$ and $J^+ = \{j \in \mathbb{N} : a_j \notin H^-\}$ it follows that

$$\hat{h}(x) = \sum_{j \in J^+} \frac{u_j(x)}{\gamma_j\, 2^j} - \sum_{j \in J^-} \frac{u_j(x)}{\gamma_j\, 2^j}, \quad x \in \mathbb{R}^2,$$

which shows that $\hat{h}$ is infinitely differentiable.

Now, with ω satisfying (i)-(v) in Lemma 7.5 the function f defined by

$$f(\lambda, x) = x^2\, \hat{h}(\lambda, x) + \omega(\lambda, x), \quad (\lambda, x) \in \mathbb{R}^2, \tag{7.6}$$

is locally Lipschitz continuous and infinitely differentiable on $\mathbb{R}^2 \setminus \{(0, 0)\}$. Since $F \subset K$ and ω is zero on K it follows that $(\lambda, x) \in F \setminus \{(0, 0)\}$ implies that $f(\lambda, x) = 0$, $x \neq 0$.

For the converse, suppose $f(\lambda, x) = 0$ and $x \neq 0$. If $\omega(\lambda, x) = 0$ then $x^2 \hat{h}(\lambda, x) = 0$, $x \neq 0$, whence $(\lambda, x) \in F \setminus \{(0, 0)\}$. Otherwise, if $\omega(\lambda, x) \neq 0$, property (i) of ω implies that (λ, x) belongs to the cone

$$\{(\lambda, x) : 0 < |x| < |\lambda|/2\} \subset (H^+ \cup H^-) \setminus K$$

which, by property (ii) implies that both $\hat{h}(\lambda, x)$ and $\omega(\lambda, x)$ have the same sign as λ. Therefore both are zero which contradicts $\omega(\lambda, x) \neq 0$. Hence $\omega(\lambda, x) \neq 0$ does not occur and the theorem is proved.   □

The following result is implicit in [48].

Theorem 7.8 *For a set F satisfying* (†) *in Definition 7.3 and the corresponding function f in Theorem 7.7, the function $R : \mathbb{R}^2 \to \mathbb{R}$ defined by* (7.3) *is continuously differentiable on $\mathbb{R}^2$, infinitely differentiable on $\mathbb{R}^2 \setminus \{(1, 0)\}$, and satisfies* (1.1b). *Moreover, (λ, x) is a non-trivial solution of* (7.1) *if and only if*

$$(\lambda, x) \in \left(F \setminus \{(0, 0)\} \right) + (1, 0). \tag{7.7}$$

If F contains no non-trivial paths, there are no non-trivial paths of non-trivial solutions of (7.1) *with this choice of R.*

Proof By Remark 7.6 and definition (7.6), xf is continuously differentiable on $\mathbb{R}^2$. Therefore the function R defined in (7.3) is continuously differentiable on $\mathbb{R}^2$, and the rest of the theorem follows from Theorem 7.1.   □

7.3   Justification of Example A

Because of Theorem 7.8 it suffices to find a closed, unbounded, connected $F \subset \mathbb{R} \times \mathbb{R}$ satisfying (†) in Definition 7.3 with,

(i) $F \subset \{(\lambda, x) \in [-\frac{1}{4}, \frac{1}{4}] \times \mathbb{R} : |\lambda| \leqslant |x|/4\}$;
(ii) $\{(0, 0)\} = F \cap \left(\mathbb{R} \times \{0\} \right)$ and $F + (0, k\pi) = F$ for all $k \in \mathbb{Z}$;
(iii) F is connected and contains no non-trivial paths.

Let $\mathscr{E} \subset \mathbb{R}^2$ be a non-degenerate, hereditarily indecomposable continuum, which exists by Theorem 6.11 and, translating and re-scaling if necessary, suppose

$$\mathscr{E} \subset [-\tfrac{1}{4}, \tfrac{1}{4}] \times [0, \pi], \quad \mathscr{E} \cap \left([-\tfrac{1}{4}, \tfrac{1}{4}] \times \{0\}\right) \neq \emptyset \text{ and } \mathscr{E} \cap \left([-\tfrac{1}{4}, \tfrac{1}{4}] \times \{\pi\}\right) \neq \emptyset.$$

Then let $\mathscr{E}_0 = \left\{ (t \sin s, s) \in \mathbb{R}^2 : (t, s) \in \mathscr{E} \right\} \subset [-\tfrac{1}{4}, \tfrac{1}{4}] \times [0, \pi]$ so that

$$\mathscr{E}_0 \cap \left([-\tfrac{1}{4}, \tfrac{1}{4}] \times \{0\}\right) = \{(0, 0)\}, \quad \mathscr{E}_0 \cap \left([-\tfrac{1}{4}, \tfrac{1}{4}] \times \{\pi\}\right) = \{(0, \pi)\}.$$

Hence, if $\mathscr{E}_0$ contains a non-trivial path, it must contain a non-trivial path in $[-\frac{1}{4}, \frac{1}{4}] \times (0, \pi)$ and, because $(t, s) \mapsto (t \sin s, s)$ is injective on $[-\frac{1}{4}, \frac{1}{4}] \times (0, \pi)$ with continuous inverse, this implies that $\mathscr{E}$ has a non-trivial path in $[-\frac{1}{4}, \frac{1}{4}] \times (0, \pi)$. But this is false by Lemma 6.10, since $\mathscr{E}$ is hereditarily indecomposable. Therefore $\mathscr{E}_0$, which is a continuous image of a continuum, is a continuum with no non-trivial

Fig. 7.2 A crude illustration of the global bifurcation of solutions of (7.1) in which there are no non-trivial paths in Example A. In it $\mathscr{P}_k = \mathscr{E}_k + (1,0)$, $k \in \mathbb{Z}$, and the dot at $\mathscr{P}_0 \cap \mathscr{P}_{-1}$ is the bifurcation point $(\lambda, x) = (1, 0)$. The construction can be adjusted so that, for a different function R, this bifurcating solution set is rotated about $(1,0)$ and its projection on the λ-axis is the whole real line

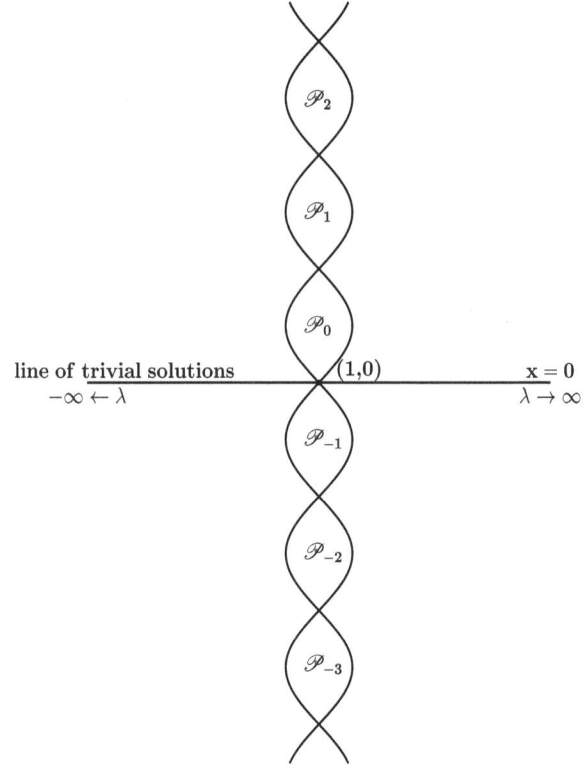

line of trivial solutions
$-\infty \leftarrow \lambda$

(1,0)

x = 0
$\lambda \to \infty$

paths. Now let

$$F = \cup_{k \in \mathbb{Z}} \mathscr{E}_k \text{ where } \mathscr{E}_k = \mathscr{E}_0 + (0, k\pi). \tag{7.8}$$

From the definitions, $(0,0) \in F$ and $F \subset \left[-\frac{1}{4}, \frac{1}{4}\right] \times \mathbb{R}$ is unbounded. Since $\mathscr{E}_0 = \{(t \sin s, s) \in \mathbb{R}^2 : (t, s) \in \mathscr{E}\}$ and $|t| \leq \frac{1}{4}$, F lies in a cone $\{(\lambda, x) \in \left[-\frac{1}{4}, \frac{1}{4}\right] \times \mathbb{R} : |\lambda| \leq |x|/4\}$. Moreover F is closed, and connected because $\mathscr{E}_k$ is connected, and $\mathscr{E}_k \cap \mathscr{E}_{k+1} = \{(0, (k+1)\pi)\}$ for all k. Since each $\mathscr{E}_k$ contains no non-trivial paths and $\mathscr{E}_k \cap \mathscr{E}_{k+1}$ is a singleton, F contains no paths. Thus by construction F satisfies (i)-(iii). Clearly F satisfies (†) with, for example, $x_j^{\pm} = \pm j\pi$, $j \in \mathbb{N}$.

With this choice of the set F, by Theorem 7.8 there is C^1-function R which is infinitely differentiable on $\mathbb{R}^2 \setminus \{(1, 0)\}$ which yields a global connected set of solutions of (7.1) namely $\overline{\mathscr{C}} = \mathscr{C}^- \cup \mathscr{C}^+$ where

$$\mathscr{C}^- = \bigcup_{k \in \mathbb{N}} \mathscr{P}_{-k} \text{ and } \mathscr{C}^+ = \bigcup_{0 \leq k \in \mathbb{Z}} \mathscr{P}_k \text{ where } \mathscr{P}_k = \mathscr{E}_k + (1, 0), \ k \in \mathbb{Z}.$$

See Fig. 7.2. This is an example where the set $\overline{\mathscr{C}}$ predicted by Theorem 1.3 (a) contains no non-trivial paths. Since the mixed partial derivative $\partial_{\lambda x} R$ is not continuous at $(1, 0)$ (Remark 7.6), Theorem 1.3(b)(i) or (ii) does not apply.   □

7.4 Justification of Example B

Compared with Example A in which the hypotheses of Theorem 1.3(a) are satisfied and the function R is infinitely differentiable except at $(1, 0)$, in this example the function R will be infinitely differentiable on all of $\mathbb{R}^2$ and satisfy the hypotheses of Theorem 1.3(b).

In the notation of (7.8) let $\mathscr{E}^+ = \bigcup_{k \in \mathbb{N}} \mathscr{E}_k$ and $\mathscr{E}^- = \bigcup_{2 \leqslant k \in \mathbb{N}} \mathscr{E}_{-k}$, so that

$$(0, \pi) \in \mathscr{E}^+ \subset \mathbb{R} \times [\pi, \infty) \text{ and } (0, -\pi) \in \mathscr{E}^- \subset \mathbb{R} \times (-\infty, -\pi]$$

where $\mathscr{E}^\pm$ are closed, connected subsets of K (see (7.4)) which contain no paths and

$$(0, \pm j\pi) \in \mathscr{E}^\pm, \ j \in \mathbb{N}, \text{ where } \|(0, \pm j\pi)\| \to \infty \text{ as } 1 \leqslant j \to \infty.$$

Then let $\mathscr{E}^0$ be the open straight-line segment $\{0\} \times (-\pi, \pi)$, and let $\overline{\mathscr{E}}$ be the union $\overline{\mathscr{E}} = \mathscr{E}^- \cup \mathscr{E}^0 \cup \mathscr{E}^+$, which is a closed, connected subset of the cone K that satisfies (†) in Definition 7.3.

Now let H^- and H^+ denote the components of $\mathbb{R}^2 \setminus \overline{\mathscr{E}}$ which contains $(-1, 0)$ and $(1, 0)$, respectively and recall from Lemma 7.4 that $H^+ \cap H^- = \emptyset$. Also by Lemma 7.2 there exists a non-negative, infinitely differentiable function h on $\mathbb{R}^2$ which is zero only on the closed set $\overline{\mathscr{E}}$, and at each point of $\overline{\mathscr{E}}$ all the derivatives of h are zero. Let

$$\tilde{h}(\lambda, x) = \begin{cases} -h(\lambda, x), & (\lambda, x) \in H^-, \\ h(\lambda, x), & \text{otherwise}, \end{cases}$$

so that $\tilde{h} \geqslant 0$ on H^+. Then let $\tilde{\omega} : \mathbb{R} \to \mathbb{R}$ be an infinitely differentiable even function with $\tilde{\omega}(0) = 1$, $\tilde{\omega}$ is decreasing on $[0, \pi/2]$ and $\tilde{\omega}(x) = 0$ when $|x| \geqslant \pi/2$. Then let

$$\tilde{f}(\lambda, x) = x^2 \tilde{h}(\lambda, x) + \lambda \tilde{\omega}(x),$$

and note that $\overline{\mathscr{E}} = \{(\lambda, x) \in \mathbb{R}^2 : \tilde{f}(\lambda, x) = 0\}$. Finally let

$$\tilde{R}(\lambda, x) = x\big(\tilde{f}(\lambda - 1, x) - \tilde{f}(\lambda - 1, 0)\big), \text{ the analogue of (7.3).}$$

Therefore $\tilde{R}$ is infinitely differentiable and by Theorem 7.8

$$\overline{\mathscr{C}} := \mathscr{C}^+ \cup \mathscr{L} \cup \mathscr{C}^- \quad \text{where} \quad \mathscr{C}^{\pm} = \mathscr{E}^{\pm} + (1, 0) \quad \text{and} \quad \mathscr{L} = \mathscr{E}^0 + (1, 0)$$

is the closure of the set of non-trivial solutions of the equation $x = \lambda x + \tilde{R}(\lambda, x)$ predicted by Theorem 1.3 (a). Furthermore, $\mathscr{L}$ exemplifies the local bifurcation result of Theorem 1.3 (b)(ii), but the only path-components of $\overline{\mathscr{C}}$ are the singletons in $\overline{\mathscr{C}} \setminus \mathscr{L}$ and the straight-line segment $\overline{\mathscr{L}} = \{1\} \times [-\pi, \pi]$.   □

7.5 Justification of Example C

In this Example the sets $\mathscr{E}^0$ and $\mathscr{E}^{\pm}$ in Example B are replaced by

$$\mathscr{E}^0 = \{(\lambda, x) : \lambda = 0, \ x \in (-\tfrac{1}{2}, \tfrac{1}{2})\},$$

$$\mathscr{E}^+ = \left\{(\lambda, x) : \lambda = 0, \ x > \tfrac{1}{2}\right\} \cup \left\{(\lambda, x + \tfrac{1}{2}) \in R^2 : (x, \lambda) \in \mathscr{B}\right\},$$

$$\mathscr{E}^- = \left\{(\lambda, x) : \lambda = 0, \ x < -\tfrac{1}{2}\right\} \cup \left\{(\lambda, x - \tfrac{1}{2} - \sigma) \in R^2 : (-x, \lambda) \in \mathscr{D}\right\},$$

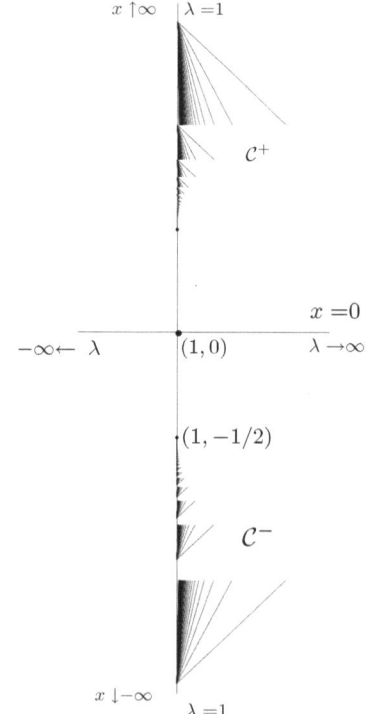

Fig. 7.3 For the two boundary points of the set $Q = \{1\} \times (-\tfrac{1}{2}, \tfrac{1}{2})$, the situation is described in (β)(ii) of Sect. 5.4, but only the set $\{(1, -\tfrac{1}{2})\}$ is a limit of a sequence of (disjoint) components of congestion points of $\mathscr{M}$ with diameters tending to zero

where $\mathscr{B}$ is the infinite broom in Fig. 5.1 and $\mathscr{D} = \bigcup_{k=0}^{\infty} D_k$ as in Fig. 5.3, Example 5.14 (II), with $\sigma > 0$. If

$$Q = \mathscr{E}^0 + (1,0), \quad \mathscr{C}^+ = \mathscr{E}^+ + (1,0) \quad \text{and} \quad \mathscr{C}^- = \mathscr{E}^- + (1,0),$$

then as in Example B the set

$$\mathscr{M} := Q \cup \mathscr{C}^+ \cup \mathscr{C}^- = \left(\mathscr{E}^+ \cup \mathscr{E}^0 \cup \mathscr{E}^- \right) + (1,0) \text{ in Fig. 7.3}$$

is the set of solutions of Eq. (7.1) for some C^∞-function R, and Q is the component of $\mathscr{M} \setminus \overline{\mathscr{N}(\mathscr{M})}$ (Definition 5.10) which contains $(1,0)$ and

$$(\partial_{\mathscr{M}} Q) \cap \mathscr{N}(\mathscr{M}) = \emptyset \quad \text{and} \quad (\partial_{\mathscr{M}} Q) \cap (\partial_{\mathscr{M}} \mathscr{N}(\mathscr{M})) = \{(1, \tfrac{1}{2}), (1, -\tfrac{1}{2})\}.$$

Thus both possibilities in Sect. 5.4 (β)(ii) occur in the this example, but at different points, $(1, \pm\tfrac{1}{2}) \in (\partial_{\mathscr{M}} Q) \setminus \mathscr{N}(\mathscr{M})$ in Fig. 7.3.

References

1. Ayala, R., Chávez, M.J., Quintero, A.: A Hahn-Mazurkiewicz theorem for generalized Peano continua. Arch. Math. **71**(4), 325–330 (1998)
2. Bartoš, A., Kubiś, W.: Hereditarily indecomposable continua as generic mathematical structures. https://arxiv.org/pdf/2208.06886.pdf
3. Bing, R.H.: A homogeneous indecomposable plane continuum. Duke Math. J. **15**, 729–742 (1948)
4. Bing, R.H.: Concerning hereditarily indecomposable continua. Pacific J. Math. **1**, 43–51 (1951)
5. Brezis, H.: Functional Analysis. Sobolev Spaces and Partial Differential Equations. Universitext. Springer, New York (2011)
6. Brouwer, L.E.J.: Zur analysis situs. Math. Ann. **68**(3), 422–434 (1910)
7. Buffoni, B., Toland, J.: Analytic Theory of Global Bifurcation. An Introduction. Princeton University Press, Princeton (2003)
8. Buffoni, B., Dancer, E.N., Toland, J.F.: The sub-harmonic bifurcation of Stokes waves. Arch. Rational Mech. Anal. **152**, 241–271 (2000)
9. Chow, S.-N., Hale, J.K.: Methods of Bifurcation Theory. Grundlehren der Mathematischen Wissenschaften, vol. 251. Springer, New York (1982)
10. Crandall, M.G., Rabinowitz, P.H.: Bifurcation from simple eigenvalues. J. Funct. Anal. **8**, 321–340 (1971)
11. Dai, G.: Bifurcation and one-sign solutions of the p-Laplacian involving a nonlinearity with zeros. Discrete Contin. Dyn. Syst. **36**, 5323–5345 (2016)
12. Dai, G.: Two Whyburn type topological theorems and its applications to Monge–Ampère equations. Calc. Var. **55**(4), Art. 97 (2016)
13. Dai, G.: Generalized limit theorem and bifurcation for problems with Pucci's operator. Topol. Methods Nonlinear Anal. **56**(1), 229–261 (2020)
14. Dancer, E.N.: Bifurcation theory for analytic operators. Proc. Lond. Math. Soc. (3) **26**, 359–384 (1973)
15. Dancer, E.N.: Global structure of the solutions of non-linear real analytic eigenvalue problems. Proc. Lond. Math. Soc. (3) **27**, 747–765 (1973)
16. Dancer, E.N.: Global solution branches for positive mappings. Arch. Rational Mech. Anal. **52**, 181–192 (1973)
17. Engelking, R.: General Topology. Sigma Series in Pure Mathematics, vol. 6. Helderman Verlag Berlin (1989)
18. Federer, H.: Geometric Measure Theory. Die Grundlehren der mathematischen Wissenschaften, Band 153. Springer New York, New York (1969)

19. Fraser Jr., R.B.: A new characterization of Peano continua. Comment. Math. Prace Mat. **16**, 247–248 (1972)
20. Hahn, H.: Mengentheoretische Charakterisierung der stetigen Kurve. Sitzungsberichte, vol. 123, pp. 2433–2489. Akademie der Wissenschaften, Vienna (1914)
21. Hart, K.P., van Mill, J., Pol, R.: Remarks on hereditary indecomposable continua. Topol. Proc. **25**, 179–206 (2000). Proceedings of the 15th Summer Conference on General Topology and its Applications/1st Turkish International Conference on Topology and its Applications (Oxford, OH/Istanbul, 2000)
22. Henrot, A., Pierre, M.: Variation et Optimisation de Formes. Une Analyse Géométrique. Mathématiques & Applications, No. 48. Springer, Berlin (2005)
23. Hocking, H.J., Young, G.S.: Topology. Addison-Wesley, Reading (1961). Dover, New York, 1988
24. Hoehn, L.C., Oversteegen, L.G.: A complete classification of homogeneous plane continua. Acta Math. **216**(2), 177–216 (2016)
25. Knaster, B.: Un continu dont tout sous-continu est indécomposable. Fund. Math. **3**, 247–286 (1922)
26. Krasinkiewicz, J., Minc, P.: Nonexistence of universal continua for certain classes of curves. Bull. Acad. Polon. Sci. Sér. Sci. Math. Astronom. Phys. **24**(9), 733–741 (1976)
27. Krasinkiewicz, J., Minc, P.: Mappings onto indecomposable continua. Bull. Acad. Polon. Sci. Sér. Sci. Math. Astron. Phys. **25**(7), 675–680 (1977)
28. Krasnolsel'skii, M.A.: Topological Methods in the Theory of Nonlinear Eigenvalue Problems. Pergamon Press, Oxford (1963). (English translation of original in Russian: Topologicheskiye Metody v Teorii Nelineinykh Integral'nykh Uravnenii. Gostekhteoretizdat, Moscow, 1956.)
29. Krasovskii, Yu.P.: On the theory of steady waves of finite amplitude. U.S.S.R. Comput. Math. Math. Phys. **1**(4), 996–1018 (1962). (Zh. Vych. Mat. 1: No. 5, 836–855, 1961.)
30. Leray, J., Schauder, J.: Topologie et équations fonctionnelles. Ann. Sci. École Norm. Sup. (3) **51**, 45–78 (1934)
31. Lewis W.: The pseudo-arc. Bol. Soc. Mat. Mexicana (3) **5**(1), 25–77 (1999)
32. Mazurkiewicz, S.: Sur les lignes de Jordan. Fundam. Math. **1**, 166–209 (1920)
33. Mazurkiewicz, S.: Sur les continus absolument indécomposables. Fundam. Math. **16**, 151–159 (1930)
34. Moise, E.E.: An indecomposable plane continuum which is homeomorphic to each of its nondegenerate subcontinua. Trans. Amer. Math. Soc. **63**, 581–594 (1948)
35. Munkres, J.R.: Topology, 2nd edn. Prentice Hall, Upper Saddle River (2000)
36. Nadler Jr., S.B.: A characterization of locally connected continua by hyperspace retractions. Proc. Amer. Math. Soc. **67**(1), 167–176 (1977)
37. Nadler Jr., S.B.: Continuum Theory. An Introduction. Monographs and Textbooks in Pure and Applied Mathematics, vol. 158. Marcel Dekker, New York (1992)
38. Newman, M.H.A.: Elements of the Topology of Plane Sets of Points, 2nd edn. Cambridge University Press (1951)
39. Oversteegen, L.G., Tymchatyn, E.D.: On hereditarily indecomposable compacta. In: Geometric and Algebraic Topology, Banach Center Publ., vol. 18. PWN, Warsaw (1986)
40. Rabinowitz, P.H.: Some global results for nonlinear eigenvalue problems. J. Funct. Anal. **7**, 487–513 (1971)
41. Royden, H.L.: Real Analysis. Macmillan, New York (1986)
42. Rudin, M.E.: A new proof that metric spaces are paracompact. Proc. Amer. Math. Soc. **20**, 603 (1969)
43. Spivak, M.: A Comprehensive Introduction to Differential Geometry, vol. 1, 3rd edn. Publish or Perish, Huston (1999)
44. Steen, L.A., Seebach, J.A.: Counterexamples in Topology. Dover Publications, Mineola (1995). (Reprint of original, Springer-Verlag, New York, 1978.)
45. Stein, E.M.: Singular Integral and Differentiability Properties of Functions. Princeton Mathematical Series, vol. 30. Princeton University Press, Princeton (1970)
46. Stone, A.H.: Paracompactness and product spaces. Bull. Amer. Math. Soc. **54**, 977–982 (1948)

47. Stuart, C.A.: Some bifurcation theory for k-set contractions. Proc. London Math. Soc. (3) **27**, 531–550 (1973)
48. Toland, J.F.: Path-connectedness in global bifurcation theory. Electron. Res. Arch. **29**(6), 4199–4213 (2021)
49. Whyburn, G.T.: Analytic Topology, vol. XXVIII. Amer. Math. Soc. Colloquium Publications, Providence (1942)
50. Whyburn, G.T.: Topological Analysis. Princeton University Press, Princeton (1958), (2nd revised edition, 1964)
51. Wilder, R.L.: Topology of Manifolds, vol. XXXII. Amer. Math. Soc. Colloquium Publications, New York (1949)
52. Willard, S.: General Topology. Dover Publications, Mineola (2004). (Reprint of the original, Addison-Wesley 1970)

Index